Introduction to Internet of Things (IoT)

Published 2023 by River Publishers
River Publishers
Alsbjergvej 10, 9260 Gistrup, Denmark
www.riverpublishers.com

Distributed exclusively by Routledge
605 Third Avenue, New York, NY 10017, USA
4 Park Square, Milton Park, Abingdon, Oxon OX14 4RN

Introduction to Internet of Things (IoT) / by Ahmed Banafa.

Routledge is an imprint of the Taylor & Francis Group, an informa business

ISBN 978-87-7022-445-1 (print)

ISBN 978-10-0092-264-6 (online)

ISBN 978-1-003-42624-0 (ebook master)

A Publication in the River Publishers series
RAPIDS SERIES IN COMPUTING AND INFORMATION SCIENCE AND TECHNOLOGY

Introduction to Internet of Things (IoT)

Ahmed Banafa

Professor of Engineering at San Jose State University USA
Instructor of Continuing Studies at Stanford University USA

Contents

Preface

I dedicate this book to my mother, I love you mom.

In 2020, experts estimated that up to 28 billion devices were connected to the Internet with only one third of them being computers, smartphones, smartwatches, and tablets, with the number estimated to hit 75 billion devices in 2025. The remaining two thirds will be other "devices" – sensors, terminals, household appliances, thermostats, televisions, automobiles, production machinery, urban infrastructure, and many other "things", which traditionally have not been Internet enabled.

This "Internet of Things" (IoT) represents a remarkable transformation of the way in which our world will soon interact. Much like the World Wide Web connected computers to networks, and the next evolution mobile devices connected people to the Internet and other people, IoT looks poised to interconnect devices, people, environments, virtual objects, and machines in ways that only science fiction writers could have imagined. In a nutshell the Internet of Things (IoT) is the convergence of connecting people, things, data and processes; it is transforming our lives, businesses and everything in between. This book "Introduction to the Internet of Things (IoT)" explores many aspects of the Internet of Things and explains many of the complicated principles of IoT and the new advancements in IoT including using Fog/Edge Computing, and AI .

This is book is for everyone who would like to have a good understanding of IoT and its applications and its relationship with business operations including: C-Suite executives, IT managers, marketing and sales people, lawyers, product and project managers, business specialists, students. It's not for programmers who are looking for codes or exercises on the different components of IoT.

Acknowledgment

I am grateful for all the support I received from my family during the stages of writing this book.

About the Author

Prof. Ahmed Banafa has extensive experience in research, operations and management, with a focus on IoT, Blockchain, Cybersecurity and AI. He is the recipient of Certificate of Honor from the City and County of San Francisco, Author & Artist Award 2019 of San Jose State University. He was named as No. 1 tech voice to follow, technology fortune teller and influencer by LinkedIn in 2018 by LinkedIn, his research featured on Forbes, IEEE and MIT Technology Review, and Interviewed by ABC, CBS, NBC, CNN, BBC, NPR, NHK, FOX, and Washington Post. He is a member of the MIT Technology Review Global Panel. He is the author of the books: "Secure and Smart Internet of Things (IoT) using Blockchain and Artificial Intelligence (AI)" which won 3 awards San Jose State University Author and Artist Award, One of the Best Technology Books of all Time Award, and One of the Best AI Models Books of All Time Award. His second book was "Blockchain Technology and Applications" won San Jose State University Author and Artist, One of the Best New Private Blockchain Books and used at Stanford University and other prestigious schools in the USA, and "Quantum Computing" Book forthcoming in 2023. He studied Electrical Engineering at Lehigh University, Cybersecurity at Harvard University and Digital Transformation at Massachusetts Institute of Technology (MIT).

CHAPTER

1

Internet of Things: The Third Wave?

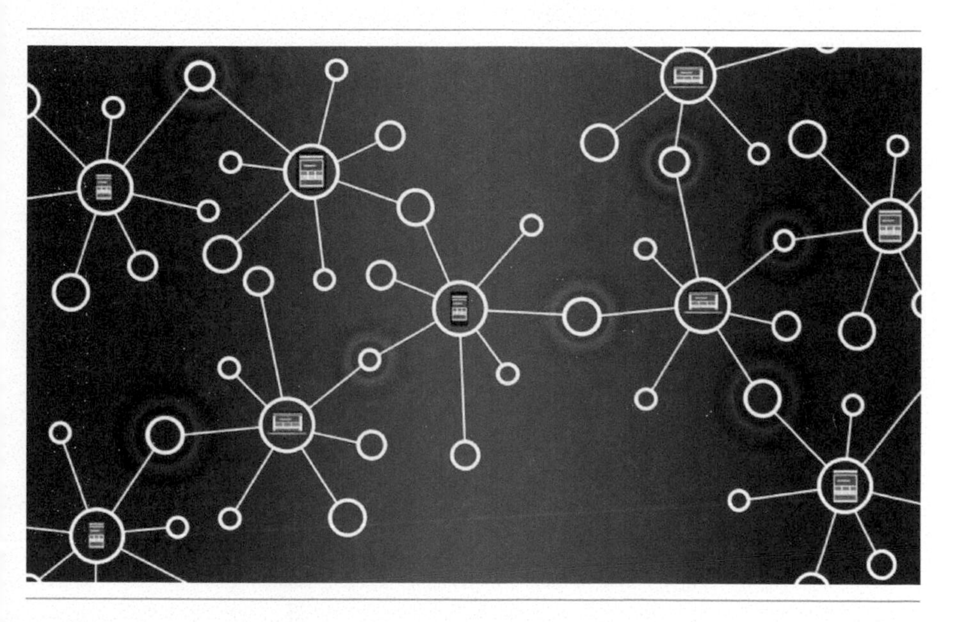

The Internet of things (IoT) is the network of physical objects accessed through the Internet. These objects contain embedded technology to interact with internal states or the external environment. In other words, when objects can sense and communicate, it changes how and where decisions are made, and who makes them, for example, Nest thermostats.

The IoT emerged as the third wave in the development of the Internet. The Internet wave of the 1990s connected 1 billion users, while the mobile wave of the 2000s connected another 2 billion. The IoT has the potential to connect 35 billion "things" to the Internet by 2022, ranging from bracelets to cars. Breakthroughs in the cost of sensors, processing power, and bandwidth to connect devices are enabling ubiquitous connections at present. Smart products like smart watches and thermostats (Nest) are already gaining traction as stated in the Goldman Sachs Global Investment Research report.

IoT has key attributes that distinguish it from the "regular" Internet, as captured by Goldman Sachs's S-E-N-S-E framework: *sensing, efficient, networked, specialized, everywhere* [1]. These attributes may change the direction of technology development and adoption, with significant implications for Tech companies – much like the transition from the fixed to the mobile Internet shifted the center of gravity from Intel to Qualcomm or from Dell to Apple.

A number of significant technology changes have come together to enable the rise of the IoT. These include the following [1]:

- **Cheap sensors**: Sensor prices have dropped to an average 60 cents from $1.30 in the past 10 years.
- **Cheap bandwidth**: The cost of bandwidth has also declined precipitously, by a factor of nearly 40 times over the past 10 years.
- **Cheap processing:** Similarly, processing costs have declined by nearly 60 times over the past 10 years, enabling more devices to be not just connected, but smart enough to know what to do with all the new data they are generating or receiving.
- **Smartphones**: Smartphones are now becoming the personal gateway to the IoT, serving as a remote control or hub for the connected home, connected car or the health and fitness devices that consumers have increasingly started to wear.
- **Ubiquitous wireless coverage:** With Wi-Fi coverage now ubiquitous, wireless connectivity is available for free or at a very low cost, given Wi-Fi utilizes an unlicensed spectrum and thus does not require monthly access fees to a carrier.
- **Big data:** As the IoT will, by definition, generate voluminous amounts of unstructured data, the availability of big data analytics is a key enabler.
- **IPv6:** Most networking equipment now supports IPv6, the newest version of the Internet Protocol (IP) standard that is intended to replace IPv4. IPv4 supports 32-bit addresses, which translates to about 4.3 billion addresses – a number that has become largely exhausted by all the connected devices globally. In contrast, IPv6 can support 128-bit addresses, translating to approximately 3.4 × 1038 addresses – an almost limitless number that can amply handle all conceivable IoT devices.

1.1 Advantages and Disadvantages of IoT

Many smart devices like laptops, smart phones, and tablets communicate with each other through the use of Wi-Fi Internet technology. Transfer these technological capabilities into ordinary household gadgets like refrigerators, washing machines, microwave ovens, thermostats, door locks among others, equip these with their own computer chips and software, and get them access to the Internet, and a "smart home" now comes to life.

The IoT can only work if these gadgets and devices start interacting with each other through a networked system. The AllJoyn Open Source Project [2], a non-profit organization devoted to the adoption of the IoT, facilitates ensuring that companies like Cisco, Sharp and Panasonic are manufacturing products compatible with a networked system and to ensure that these products can interact with each other.

The advantages of these highly networked and connected devices are productive and enhanced quality of lives for people. For example, health monitoring can be rather easy with connected RX bottles and medicine cabinets. Doctors supervising patients can monitor their medicine intake as well as measure blood pressure and sugar levels and alert them online when something goes wrong with their patients.

Considering energy conservation, household appliances can suggest optimal settings based on the user's energy consumption like turning on the ideal temperature just before the owner arrives home as well as turning on and off the lights whenever the owner is out on vacation just to create the impression that somebody is still left inside the house to prevent burglars from attempting to enter.

Smart refrigerators, on the other hand, can suggest food supplies that are low on inventory and needs immediate replenishment. The suggestions are based on the user's historical purchasing behavior and trends. Wearable technology is also part of the IoT, as these devices can monitor sleeping patterns, workout measurements, sugar levels, blood pressure and connecting these data to the user's social media accounts for tracking purposes. Figure 1.1 summarizes all four components of IoT with examples of each component.

The most important disadvantage of the IoT is with regard to privacy and security issues. Smart home devices have the ability to devour a lot of data and information about a user. These data can include personal schedules, shopping habits, medicine intake schedules, and even location of the user at any given time. If these data are misused, great harm and damage can be done to people.

The other disadvantage is the fact that most devices are not yet ready to communicate with another brand of device. Specific products can only be networked with their fellow products under the same brand name. It is good that The AllJoyn Open Source Project [2], is ensuring connectivity, but the reality

Figure 1.1: IoT components.

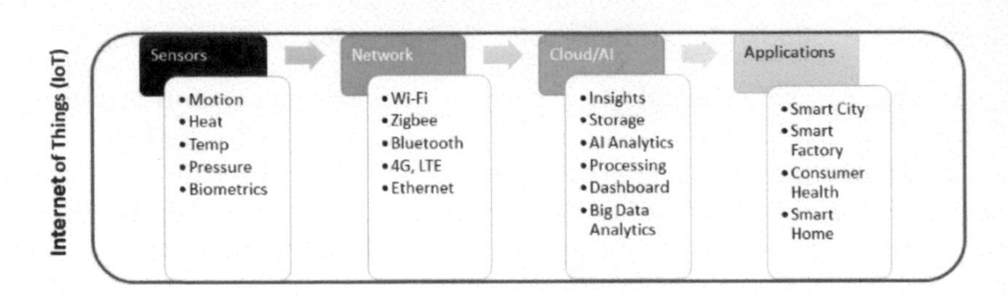

of a “universal remote control” for all these devices and products is still in its infancy.

CHAPTER

2

The Industrial Internet of Things (IIoT): Challenges, Requirements, and Benefits

The idea of a smarter world, where systems with sensors and local processing are connected to share information, is taking hold in every single industry. These systems will be connected on a global scale with users and each other to help them make more informed decisions. Many labels have been given to this overarching idea, but the most ubiquitous is the Internet of things (IoT). The IoT includes everything from smart homes, mobile fitness devices, and connected toys to the industrial Internet of things (IIoT) with smart agriculture, smart cities, smart factories, and the smart grid.

The IIoT is a network of physical objects, systems, platforms and applications that contain embedded technology to communicate and share intelligence with each other, the external environment and with people. The adoption of the IIoT is being enabled by the improved availability and affordability of sensors, processors, and other technologies that have helped facilitate capture of, and access to, real-time information.

The IIoT can be characterized as a vast number of connected industrial systems that are communicating and coordinating their data analytics and actions to improve industrial performance and benefit society as a whole. Industrial systems that interface the digital world to the physical world through sensors and actuators that solve complex control problems are commonly known as cyber-physical systems. These systems are being combined with big data solutions to gain deeper insight through data and analytics.

Imagine industrial systems that can adjust to their own environments or even their own health. Instead of running to failure, machines would schedule their own maintenance or, better yet, adjust their control algorithms dynamically to compensate for a worn part and then communicate those data to other machines and the people who rely on those machines. By making machines smarter through local processing and communication, the IIoT could solve problems in ways that were previously inconceivable. But, as the saying goes, "If it was easy, everyone would be doing it." As innovation grows, so does the complexity, which makes the IIoT face a significant challenge that no company alone can meet.

At its root, the IIoT is a vast number of connected industrial systems that communicate and coordinate their data analytics and actions to improve performance and efficiency and reduce or eliminate downtime. A classic example is industrial equipment on a factory floor that can detect minute changes in its operations, determine the probability of a component failure, and then schedule maintenance of that component before its failure can cause unplanned downtime that could cost millions of dollars.

The possibilities in the industrial space are nearly limitless: smarter and more efficient factories, greener energy generation, self-regulating buildings that optimize energy consumption, cities that adjust and can adjust traffic patterns to respond to congestion and more. But, of course, implementation will be a challenge.

2.1 IIoT, IoT, and M2M

The main difference between IoT and IIoT is that consumer IoT often focuses on convenience for individual consumers, while industrial IoT is strongly focused on improving the efficiency, safety, and productivity of operations with a focus on return on investment. M2M is a subset of IIoT, which tends to focus very specifically on machine-to-machine communications, where IoT expands that to include machine-to-object/people/infrastructure communications. The IIoT is about making machines more efficient and easier to monitor.

2.1.1 IIoT Challenges (Figure 2.1)

- Precision
- Adaptability and scalability
- Security
- Maintenance and updates
- Flexibility.

Figure 2.1: IIoT challenges.

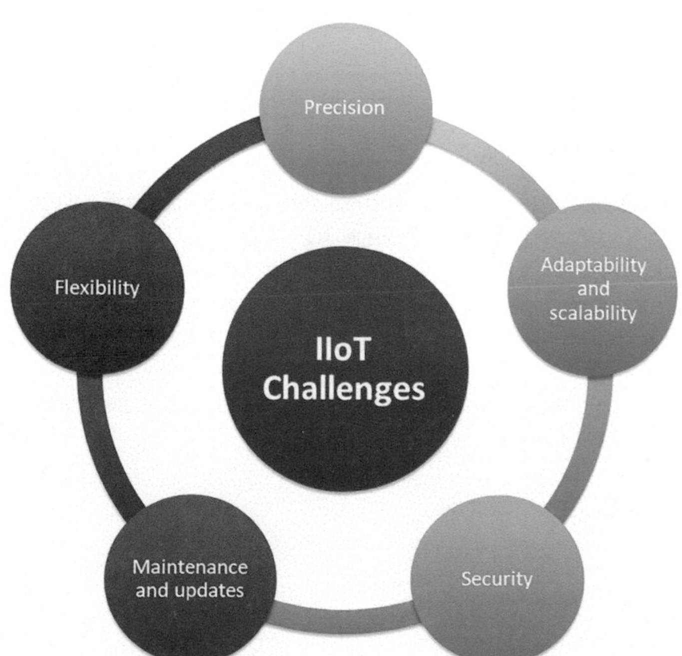

2.1.2 IIoT Requirements (Figure 2.2)

- Cloud computing
- Access (anywhere, anytime)
- Security
- Big data analytics
- UX (user experience)
- Assets management
- Smart machines.

Figure 2.2: IIoT requirements.

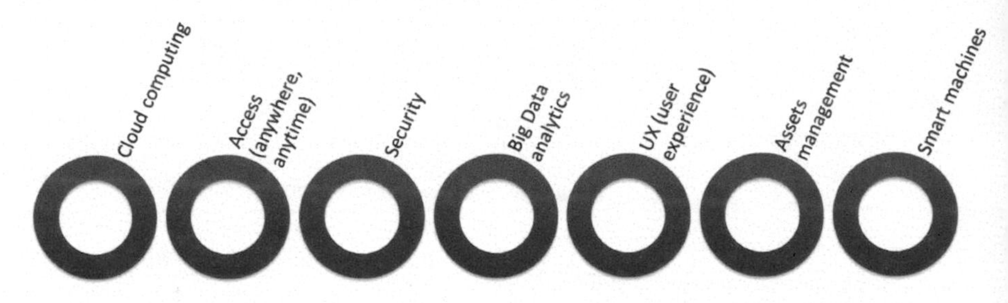

2.1.3 IIoT Benefits (Figure 2.3)

- Vastly improved operational efficiency (e.g., improved uptime, asset utilization) through predictive maintenance and remote Management.
- The emergence of an outcome economy, fueled by software- driven services; innovations in hardware; and the increased visibility into products, processes, customers and partners.
- New connected ecosystems, coalescing around software platforms that blur traditional industry boundaries.
- Collaboration between humans and machines, which will result in unprecedented levels of productivity and more engaging work Experiences.

Figure 2.3: IIoT benefits.

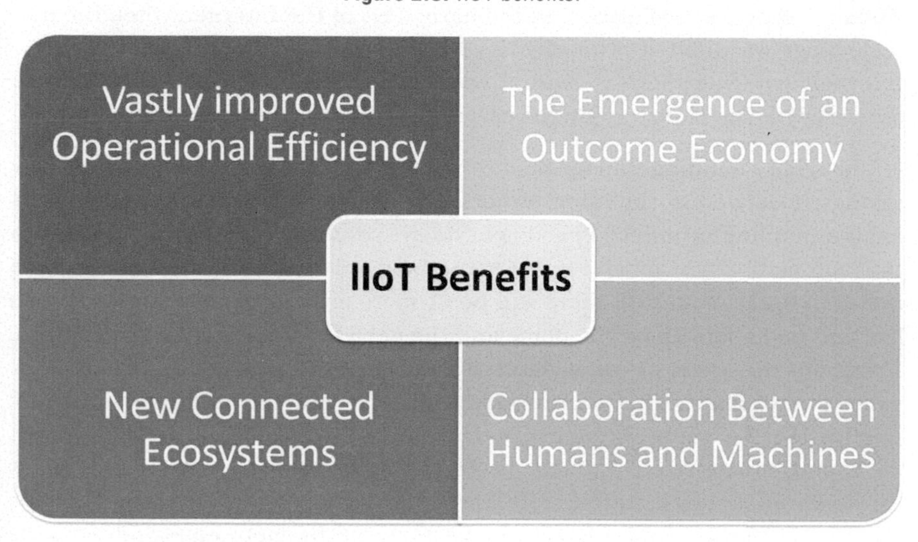

2.2 The Future of the Industrial Internet of Things

Accenture estimates that it could add more than $10 trillion to the global economy by 2030 [3]. And that number could be even higher if companies were to take bolder actions and make greater investments in innovation and technology than they are doing today.

The good news is the IIoT is already here, at least among the most forward-thinking companies. The challenge is that most businesses are not ready to take the plunge. According to an Accenture survey of more than 1400 business leaders, only one-third (36%) claim they fully grasp the implications of the IIoT. Just 7% have developed a comprehensive IIoT strategy with investments to match.

One of the reasons is the as-yet limited ability to leverage machine intelligence to do more than enhance efficiencies on the factory floor and evolve to create entirely new value-added services, business models and revenue streams.

So far, businesses have made progress in applying the IIoT to reduce operational expenses, boost productivity or improve worker safety. Drones, for example, are being used to monitor remote pipelines, and intelligent drilling equipment can improve productivity in mines. Although these applications are valuable, they are reminiscent of the early days of the Internet, when the new technology was limited primarily to speeding up work processes. As with the Internet, however, there is more growth, innovation, and value that can be derived with smart IIoT applications.

Imagine a building management company charging fees based on the energy savings it delivers to building owners. Or an airline company rewarding its engine supplier for reduced passenger delays resulting from performance data that automatically schedules maintenance and orders spare parts while a plane is still in flight. With IIoT, there will be no more missing planes; information is live and up-to-date about the plane and the need for a black box will diminish. These are the kinds of product–service hybrid models that can provide new value to customers. This transformation in business will also have dramatic implications for the workforce. Clearly, the IIoT will digitize some jobs that have, until now, resisted automation. However, the vast majority of executives we surveyed believe that the IIoT will be a net creator of jobs. Perhaps more importantly, routine tasks will be replaced by more engaging work, as technology allows workers to do more. As the focus shifts from products to customers, knowledge-intensive work will be required to handle exceptions and tailor solutions. Virtual teams will be able to collaborate, creating and experimenting in more spontaneous and responsive environments.

The transformation in business models draws a parallel with those sparked by the emergence of electricity. It took decades to move from lighting streets to creating the electric grid. The mass assembly line soon became commonplace, requiring an entirely new set of skills, management approaches, and factory design. The United States was the first country to seize that opportunity and create an economy-wide impact with electricity. That helped the nation develop and lead subsequent innovations that became entirely new sectors: domestic appliances, the semiconductor industry, software and the Internet itself.

CHAPTER

3

Internet of Things: Security, Privacy, and Safety

The Internet of things (IoT) presents numerous benefits to consumers, and has the potential to change the ways in which consumers interact with technology in fundamental ways. In the future, the IoT is likely to meld the virtual and physical worlds together in ways that are currently difficult to comprehend. From a security and privacy perspective, the predicted pervasive introduction of sensors and devices into currently intimate spaces – such as the home, the car, and with wearables and ingestible, even the body – poses particular challenges. As physical objects in our everyday lives increasingly detect and share observations about us, consumers will likely continue to want privacy.

IoT devices are poised to become more pervasive in our lives than mobile phones and will have access to the most sensitive personal data, such as social security numbers and banking information, as their numbers are exponentially multiplied. A couple of security concerns on a single device such as a mobile phone can quickly turn to 50 or 60 concerns when considering multiple IoT devices in an interconnected home or business. In light of the importance of what IoT devices have access to, it is important to understand their security risk.

The Federal Trade Commission [4], has warned recently that the small size and limited processing power of many connected devices could limit the use of encryption and other security measures; it may also be difficult to patch flaws in low-cost and essentially disposable IoT devices.

The growth in these connected devices will spike over the next several years, according to numbers accumulated by Cisco Systems. What Cisco officials call the Internet of everything [5] will generate $19 trillion [6] in new revenues for businesses worldwide in the coming few years, and IDC analysts expect the IoT technology and services market to hit $8.9 trillion by the end of the decade.

However, while it may prove a financial boon for businesses and meet consumers' insatiable desire for more devices, the IoT will also increase the potential attack surface for hackers and other cyber-criminals. More devices online means more devices that need protecting [7], and IoT systems are not usually designed for cybersecurity. The sophistication of cybercriminals is increasing, and the data breaches that are becoming increasingly familiar will only continue.

Internet of things security is no longer a foggy future issue, as an increasing number of such devices enter the market, and our lives, from self-parking cars to home automation systems to wearable smart devices. Google CEO Eric Schmidt [8] told world leaders at the World Economic Forum in Davos, Switzerland "There will be so many sensors, so many devices, that you won't even sense it, it will be all around you. It will be part of your presence all the time."

Issues around mobile security are already a challenge in this era of always connected devices. Think how much greater those challenges will be if a business has, for example, 10 IoT connected devices. And it is not going to get any easier. As the IoT evolves, there will be billions of connected devices – and each one represents a potential doorway into your IT infrastructure and your company or personal data.

3.1 IoT's Threats

We can list the threats of IoT under three categories: privacy, security, and safety. Experts say the security threats of the IoT are broad and potentially even crippling to systems. Since the IoT will have critical infrastructure components, it presents a good target for national and industrial espionage, as well as denial of service and other attacks. Another major area of concern is the privacy of personal information that will potentially reside on networks, also a likely target for cybercriminals.

One thing to keep in mind when evaluating security needs is that the IoT is still very much a work in progress. Many things are connected to the Internet now [9], and we will see an increase in this and the advent of contextual data sharing and autonomous machine actions based on that information, the IoT is the allocation of a virtual presence to a physical object, as it develops, and these virtual presences will begin to interact and exchange contextual information, and the devices will make decisions based on this contextual device. This will lead to very physical threats around national infrastructure, possessions (e.g., cars and homes), environment, power, water, food supply, etc.

As a variety of objects become part of an interconnected environment, we have to consider that these devices have lost physical security, as they are going to be located in inhospitable environments, instantly accessible by the individual who is most motivated to tamper with the controls, attackers could potentially intercept, read or change data; they could tamper with control systems and change functionality, all adding to the risk scenarios.

3.2 Threats are Real

Among the recent examples, one involves researchers who hacked into two cars and wirelessly disabled the brakes, turned the lights off and switched the brakes full on – all beyond the control of the driver. In another case, a luxury yacht was lured off course by researchers hacking the GPS signal that it was using for navigation.

Home control hubs have been found to be vulnerable, allowing attackers to tamper with heating, lighting, power, and door locks, and other cases involve industrial control systems being hacked via their wireless network and sensors.

We are already seeing hacked TV sets, video cameras, and child monitors which have raised privacy concerns. Even power meters have been hacked, which to date have been used to steal electric power, adds Paul Henry, a principal at security consulting firm vNet Security LLC in Boynton Beach, FL, USA, and a senior instructor at the SANS Institute, a cooperative research and education organization in Bethesda, MD, USA. "A recent article spoke of a 'hacked light bulb"' Henry says. "I can imagine a worm that would compromise large numbers of these Internet-connected devices and amass them in to a botnet of some kind. Remember it is not just the value or power of the device that the bad guy wants; it is the bandwidth it can access and use in a DDoS [distributed denial-of-service] attack."

The biggest concern, Henry says, is that the users of IoT devices will not regard the security of the devices they are connecting as being of great concern. "The issue is that the bandwidth of a compromised device can be used to attack a third party," he says. "Imagine a botnet of 100,000,000 IoT devices all making legitimate Web site requests on your corporate Web site at the same time."

Experts say the IoT will likely create unique and, in some cases, complex security challenges for organizations. As machines become autonomous, they are able to interact with other machines and make decisions that have an impact on the physical world. We have seen problems with automatic trading software, which can get trapped in a loop causing market drops. The systems may have failsafe built in, but these are coded by humans who are fallible, especially when they are writing code that works at the speed and frequency that computer programs can operate.

If a power system were hacked and the lights were turned off in an area of a city. No big deal perhaps for many, but for the thousands of people in subway stations hundreds of feet underground in pitch darkness, the difference is massive. IoT allows the virtual world to interact with the physical world and that brings big safety issues.

3.3 What Can We Do

While threats will always exist with the IoT as they do with other technology endeavors, it is possible to bolster the security of IoT environments using security tools such as data encryption, strong user authentication, resilient coding, and standardized and tested APIs that react in a predictable manner.

Some security tools will need to be applied directly to the connected devices. “The IoT and its cousin BYOD have the same security issues as traditional computers”, says Randy Marchany, CISO at Virginia Tech University and the director of Virginia Tech’s IT Security Laboratory [10]. “However, IoT devices usually don’t have the capability to defend themselves and might have to rely on separate devices such as firewalls [and] intrusion detection/prevention systems. Creating a separate network segment is one option.” “In fact, the lack of security tools on the devices themselves or a lack of timely security updates on the devices is what could make securing the IoT somewhat more difficult from other types of security initiatives”, Marchany says. “Physical security is probably more of an issue, since these devices are usually out in the open or in remote locations and anyone can get physical access to it”, says Marchany. “Once someone has physical access to the device, the security concerns rise dramatically.”

“It doesn’t help that vendors providing IoT technologies most likely have not designed security into their devices”, says Marchany. “In the long term, IT executives should start requiring the vendors to assert [that] their products aren’t vulnerable to common attacks such as those listed in the OWASP [11] [Open Web Application Security Project] Top 10 Web Vulnerabilities”, he says. IT and security executives should “require vendors to list the vulnerabilities they know exist on their devices as part of the purchase process”.

Security needs to be built in as the foundation of IoT systems, with rigorous validity checks, authentication and data verification, and all the data need to be encrypted. At the application level, software development organizations need to be better at writing code that is stable, resilient and trustworthy, with better code development standards, training, threat analysis and testing. As systems interact with each other, it is essential to have an agreed interoperability standard, which is safe and valid. Without a solid bottom–top structure, we will create more threats with every device added to the IoT. What we need is a secure and safe IoT with privacy-protected; a tough trade-off but not impossible.

CHAPTER

4

Internet of Things: More than Smart "Things"

Experts forecast that by 2022, up to 35 billion devices will be connected to the Internet with only one-third of them being computers, smartphones and tablets. The remaining two-thirds will be other "devices" – sensors, terminals, household appliances, thermostats, televisions, automobiles, production machinery, urban infrastructure and many other "things", which traditionally have not been Internet- enabled.

This "Internet of things" (IoT) represents a remarkable transformation in the way in which our world will soon interact. Much like the World Wide Web connected computers to networks, and the next evolution connected people to the Internet and other people, IoT looks poised to interconnect devices, people, environments, virtual objects, and machines in ways that only science fiction writers could have imagined.

In a nutshell, the IoT is the convergence of connecting people, things, data and processes in transforming our lives, business, and everything in between.

A fair question to ask at this point is how IoT differs from machine to machine (M2M), which has been around for decades. Is IoT simply M2M with IPv6 addresses or is it really something revolutionary?

To answer this question, you need to know that M2M, built on proprietary and closed systems, was designed to move data securely in real time and mainly used for automation, instrumentation, and control. It was targeted at point solutions (e.g., using sensors to monitor an oil well), deployed by technology buyers and seldom integrated with enterprise applications to help improve corporate performance.

By contrast, IoT is built with interoperability in mind and is aimed at integrating sensor/device data with analytics and enterprise applications to provide unprecedented insights into business processes, operations, and supplier and customer relationships. IoT is therefore a "tool" that is likely to become invaluable to CEOs, CFOs, and general managers of business units.

The technical definition of the IoT is a network of physical objects accessed through the Internet. These objects contain embedded technology to interact with internal states or the external environment.

In other words, when an object can sense and communicate, it changes how and where decisions are made and who makes them.

Due to the great breadth in the number of industries that have begun to be or soon will be affected by IoT, it is not right to define IoT as a unified "market". Rather, in an abstract sense, it is a technology "wave" that will sweep across many industries at different points in time. The IoT is emerging as the third wave in the development of the Internet. The Internet wave of the 1990s connected 1 billion users, while the mobile wave of the 2000s connected another 1 billion. The IoT has the potential to connect 10 times as many (35 billion) "things" to the Internet by 2022, ranging from bracelets to cars.

Breakthroughs in the cost of sensors, processing power and bandwidth to connect devices are enabling ubiquitous connections currently. Smart products like smart watches and thermostats (Nest) are already gaining traction as stated in the Goldman Sachs Global Investment Research report [1].

IoT has key attributes that distinguish it from the "regular" Internet, as captured by Goldman Sachs's S-E-N-S-E framework: *sensing, efficient, networked, specialized, everywhere.* These attributes may change the direction of technology development and adoption, with significant implications for Tech companies (Figure 4.1).

Figure 4.1: IoT key attributes.

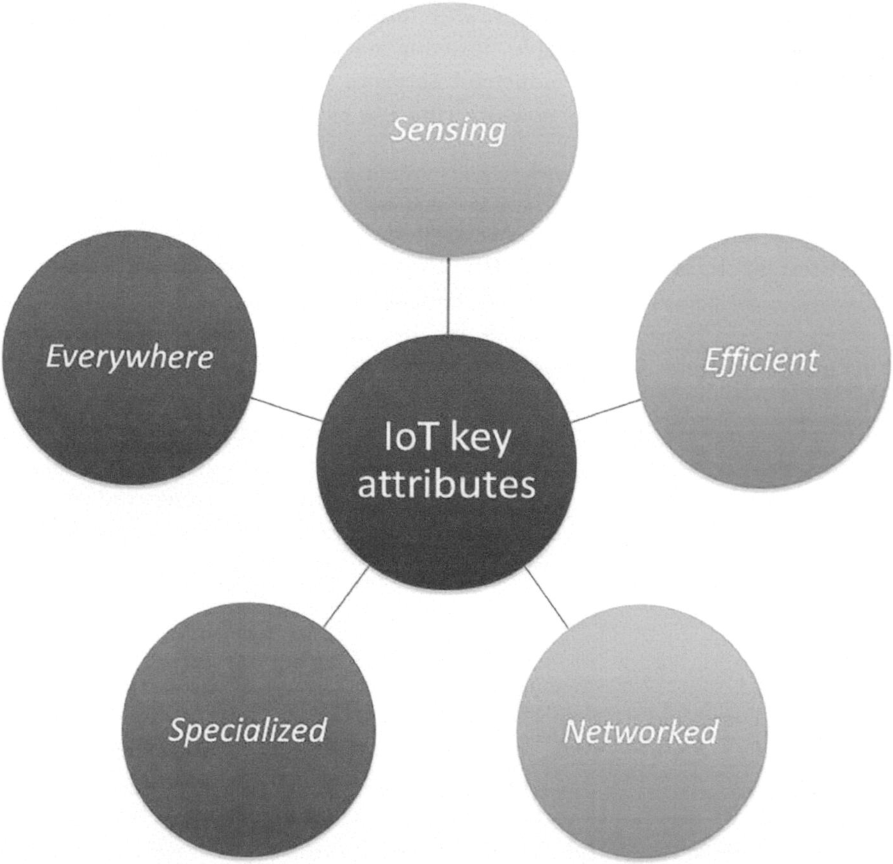

4.1 Industrial Internet

"The Internet of things will give IT managers a lot to think about", said Vernon Turner, Senior Vice President of Research at IDC. "Enterprises will have to address every IT discipline to effectively balance the deluge of data from devices that are connected to the corporate network. In addition, IoT will drive tough organizational structure changes in companies to allow innovation to be transparent to everyone, while creating new competitive business models and products."

The IoT is shaping modern business manufacturing to marketing. A lot has already been changed since the inception of the Internet, and much more will get changed with the greater Internet connectivity and reach. The global network connecting people, data and machines transforming the modern business is also called the industrial Internet [12]. The so-called industrial Internet has the potential of contributing $10 to $15 trillion to global GDP in the next two decades.

The "buzz" surrounding IoT has so far been more focused on the home, consumer, and wearables markets, and tends to overshadow the enormous potential of Internet Protocol (IP) connected products in industrial and business/enterprise worlds. IoT in the consumer world is effectively a greenfield opportunity with no installed base and no dominant vendors, whereas there are many examples of connected products in this arena. The definition of the "industrial and business/enterprise Internet" for IoT purposes refers to all non-consumer applications of the IoT, ranging from smart cities, smart power grids, connected health, retail, supply chain, and military applications. The technologies and solutions needed for creating smart connected products and processes share many common attributes across industrial and business verticals.

4.2 The IoT Value Chain

The IoT value chain is broad, extremely complex and spans many industries including those as diverse as semiconductors, industrial automation, networking, wireless and wireline operators, software vendors, security, and systems integrators. Because of this complexity, very few companies will be able to successfully solve all of the associated problems or exploit the potential opportunities.

4.3 Internet of Things Predictions

According to IDC, IoT will go through a huge growth in the coming years in many directions:

1. **IoT and the cloud:** Within the next 5 years, more than 90% of all IoT data will be hosted on service provider platforms as cloud computing reduces the complexity of supporting IoT "data blending".
2. **IoT and security:** Within the next 2 years, 90% of all IT networks will have an IoT-based security breach, although many will be considered "inconveniences". Chief information security officers (CISOs) will be forced to adopt new IoT policies.
3. **IoT at the edge:** By 2022, 53% of IoT-created data will be stored, processed, analyzed and acted upon close to, or at the edge of, the network.
4. **IoT and network capacity:** Within the next 3 years, 50% of IT networks will transition from having excess capacity to handle the additional IoT devices to being network constrained with nearly 10% of sites being overwhelmed. **IoT and non-traditional infrastructure:** By 2022, 90% of datacenter and enterprise systems management will rapidly adopt new business models to manage non-traditional infrastructure and BYOD device categories.
5. **IoT and vertical diversification:** Today, over 50% of IoT activity is centered in manufacturing, transportation, smart city, and consumer applications, but within the next 5 years, all industries will have rolled out IoT initiatives.
6. **IoT and the smart city:** Competing to build innovative and sustainable smart cities, local government will represent more than 40% of all government external spending to deploy, manage, and realize the business value of the IoT by 2022.
7. **IoT and embedded systems:** By 2022, 80% of IT solutions originally developed as proprietary, closed-industry solutions will become open-sourced, allowing a rush of vertical-driven IoT markets to form.
8. **IoT and wearables:** Within the next 5 years, 40% of wearables will have evolved into a viable consumer mass market alternative to smartphones.
9. **IoT and millennials**: By 2022, 23% of the population will be millennials and will be accelerating IoT adoption due to their reality of living in a connected world.

4.4 Challenges Facing IoT

IoT is shaping human life with greater connectivity and ultimate functionality through ubiquitous networking to the Internet. It will be more personal and predictive and merge the physical world and the virtual world to create a highly personalized and often predictive connected experience. With all the promises and potential, IoT still has to resolve three major issues [13]: unified standards for devices, privacy and security. Without the consideration of strong security at all joints of the IoT and protection of data, the progress of IoT will be hindered by litigation and social resistance. The expansion of IoT will be slow without common standards for the connected devices or sensors.

CHAPTER

5

Internet of Things: Myths and Facts

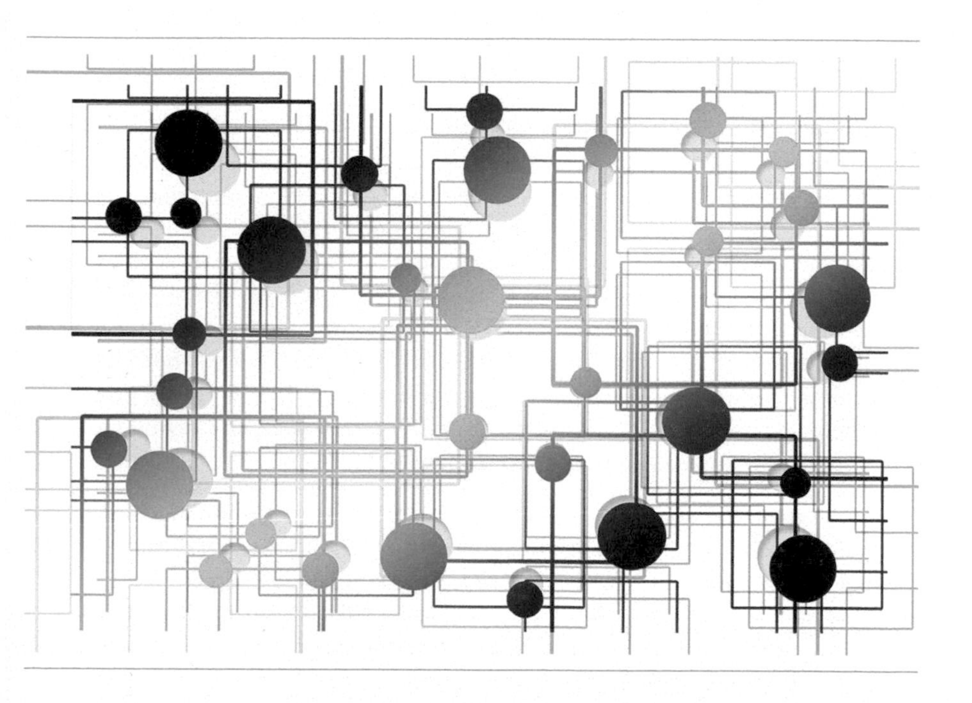

Any new technology involves a certain amount of uncertainty and business risk. In the case of the IoT, however, many of the risks have been exaggerated or misrepresented. While the IoT vision will take years to mature fully, the building blocks to begin this process are already in place. Key hardware and software are either available today or under development; stakeholders need to address security and privacy concerns, collaborate to implement the open standards that will make the IoT safe, secure, reliable and interoperable, and allow the delivery of secured services as seamlessly as possible.

The IoT is a concept that describes a totally interconnected world. It is a world where devices of every shape and size are manufactured with "smart" capabilities that allow them to communicate and interact with other devices, exchange data, make autonomous decisions and perform useful tasks based on preset conditions. It is a world where technology will make life richer, easier, safer and more comfortable.

Cisco is expecting the industry to gross over $19 trillion in the next few years. However, the problem is that these "things" have myths surrounding them, some of which are impacting how organizations develop the apps to support them.

5.1 IoT and Sensors

According to Cisco, "The fundamental problem posed by the Internet of things is that network power remains very centralized [14]. Even in the era of the cloud, when you access data and services online you're mostly communicating with a relative few massive datacenters that might not be located particularly close to you. That works when you're not accessing a ton of data and when latency isn't a problem, but it doesn't work in the Internet of things, where you could be doing something like monitoring traffic at every intersection in a city to more intelligently route cars and avoid gridlock. In that instance, if you had to wait for that chunk of data to be sent to a datacenter hundreds of miles away, processed, and then commands sent back to the streetlights, it would already be too late – the light would have already needed to change."

Cisco says that the solution is to do more computing closer to the sensors (fog/edge computing [15]) that are gathering the data in the first place, so that the amount of data that needs to be sent to the centralized servers is minimized and the latency is mitigated. Cisco says that this data-crunching capability should be put on the router. This, however, is only part of the story. Getting the right data from the right device at the right time is not just about hardware and sensors, instead it is about data intelligence. If you can understand data and only distribute what is important, at the application level, this is more powerful than any amount of hardware you throw at the problem.

This prioritization of data should be done at the application level where there is logic. Combine this with data caching at the network edge and you have a solution that reduces latency.

5.2 IoT and Mobile Data

Smartphones certainly play a role in collecting some of this data and providing a user interface for accessing IoT applications, but they are ill-suited to play a more central role. Consider the example of home automation, it hardly makes sense for critical home-monitoring and security applications, such as those that protect an elderly resident against an accident or illness, to rely upon a smartphone as their decision-making hub. What happens when that person travels and his smartphone goes into airplane mode? Does his home security get interrupted, or home electricity shut down?

Such examples make it clear that the IoT will, with a few exceptions (such as "wearable" technology and biomonitoring systems) and some automobile-related applications, rely mostly upon dedicated gateways and remote processing solutions – not on smartphones and mobile apps.

Today, without any IoT services, more than 80% of the traffic over LTE networks goes through Wi-Fi access points. What happens when that data increases by 22 times? In addition, cellular networks and communication devices have serious drawbacks in areas such as cost, power consumption, coverage and reliability.

So, will the IoT have a place for smartphones and cellular communications? Absolutely, but in terms of performance, availability, cost, bandwidth, power consumption, and other key attributes, the IoT will require a much more diverse and innovative variety of hardware, software and networking solutions.

5.3 IoT and the Volume of Data

The IoT is going to produce a lot of data – an avalanche. As a result, some IoT experts believe that we will never be able to keep up with the ever-changing and ever-growing amount of data being generated by the IoT because it is just not possible to monitor it all. Among all the data that is produced by the IoT, not all of it needs to be communicated to end-user apps such as real-time operational intelligence apps. This is because a lot of the chatter generated by devices is useless and does not represent any change in state. The apps are only interested in state changes, e.g. a light being on or off, a valve being open or shut and a traffic lane being open or closed. Rather than bombarding the apps with all of the device updates, apps should only be updated when the state changes.

5.4 IoT and Datacenters

Some argue that the datacenter is where all the magic happens for IoT. The datacenter is absolutely an important factor for the IoT; after all, this is where the data will be stored. But the myth here is that the datacenter is where the magic happens. What about the network? After all, IoT is nothing without the Internet actually supporting the distribution of information. So you might be able to store it or analyze it in a datacenter, but if the data cannot get there in the first place, is too slow in getting there or you cannot respond back in real time, then there is no IoT.

5.5 IoT is a Future Technology

The IoT is simply the logical next step in an evolutionary process. The truth is that the technological building blocks of the IoT, including microcontrollers, microprocessors, environmental and other types of sensors, and short-range and long-range networking communications, are in widespread use today. They have become far more powerful, even as they get smaller and less expensive to produce. The IoT, as we define it, while evolving the existing technologies further, simply adds one additional capability – a secured service infrastructure – to this technology mix. Such a service infrastructure will support the communication and remote control capabilities that enable a wide variety of Internet-enabled devices to work together [16].

5.6 IoT and Current Interoperability Standards

Everybody involved in the standards-making process knows that one size will not fit all – multiple (and sometimes overlapping) standards are a fact of life when dealing with evolving technology. At the same time, a natural pruning process will encourage stakeholders to standardize and focus on a smaller number of key standards. Standards issues pose a challenge, but these will be resolved as the standards process continues to evolve.

The IoT will eventually include billions of interconnected devices. It will involve manufacturers from around the world and countless product categories. All of these devices must communicate, exchange data and perform closely coordinated tasks – and they must do so without sacrificing security or performance.

This sounds like a recipe for mass confusion. Fortunately, the building blocks to accomplish many of these tasks are already in place. Global standards bodies such as IEEE, International Society of Automation (ISA), the World Wide Web Consortium (W3C), and OMA SpecWorks (to name a few) bring together manufacturers, technology vendors, policy-makers and other interested stakeholders. As a result, while standards issues pose a short-term challenge for building the IoT, the long-term process for resolving these challenges is already in place.

5.7 IoT and Privacy and Security

Security and privacy are major concerns – and addressing these concerns is a top priority. These are legitimate concerns. New technology often carries the potential for misuse and mischief, and it is vital to address the problem before it hinders personal privacy and security, innovation or economic growth. Manufacturers, standards organizations and policy-makers are already responding on several levels.

At the device level, security researchers are working on methods to protect embedded processors that, if compromised, would halt an attacker's ability to intercept data or compromise networked systems. At the network level, new security protocols will be necessary to ensure end-to-end encryption and authentication of sensitive data, and since with the IoT the stakes are higher than the Internet, the industry is looking at full system-level security and optimization.

5.8 IoT and Limited Vendors

Open platforms and standards will create a base for innovation from companies of all types and sizes:

- **Open hardware architectures**: Open platforms are a proven way for developers and vendors to build innovative hardware with limited budgets and resources.
- **Open operating systems and software:** The heterogeneous nature of the IoT will require a wide variety of software and applications, from embedded operating systems to big data analytics [17] and cross-platform development frameworks. Open software is extremely valuable in this context, since it gives developers and vendors the ability to adopt, extend and customize applications as they see fit – without onerous licensing fees or the risk of vendor lock-in.

Open standards: As we discussed earlier, open standards and interoperability are vital to building the IoT. An environment where such a wide variety of devices and applications must work together simply cannot function unless it remains free from closed, proprietary standards.

Virtually all of the vendors, developers and manufacturers involved in creating the IoT understand that open platforms will spur innovation and create rich opportunities for competition. Those that do not understand this may suffer the same fate as those that promoted proprietary networking standards during the Internet era; they were sidelined and marginalized.

5.9 Conclusion

The reality of the IoT is that if you want to distribute data from the "thing" across the network in real time over unreliable networks, you need intelligent data distribution. To lighten the load on the network by reducing your bandwidth usage, you need to understand your data. By understanding it, you can apply intelligence to only distribute what is relevant or what has changed. This means you send only small pieces of data across a congested network. The result is IoT apps with accurate, up-to date information, at scale, because you will be able to cope with the millions of devices connecting to your back end. You will not be hit with huge pieces of data at once, shutting down your services.

CHAPTER

6

Three Major Challenges Facing IoT

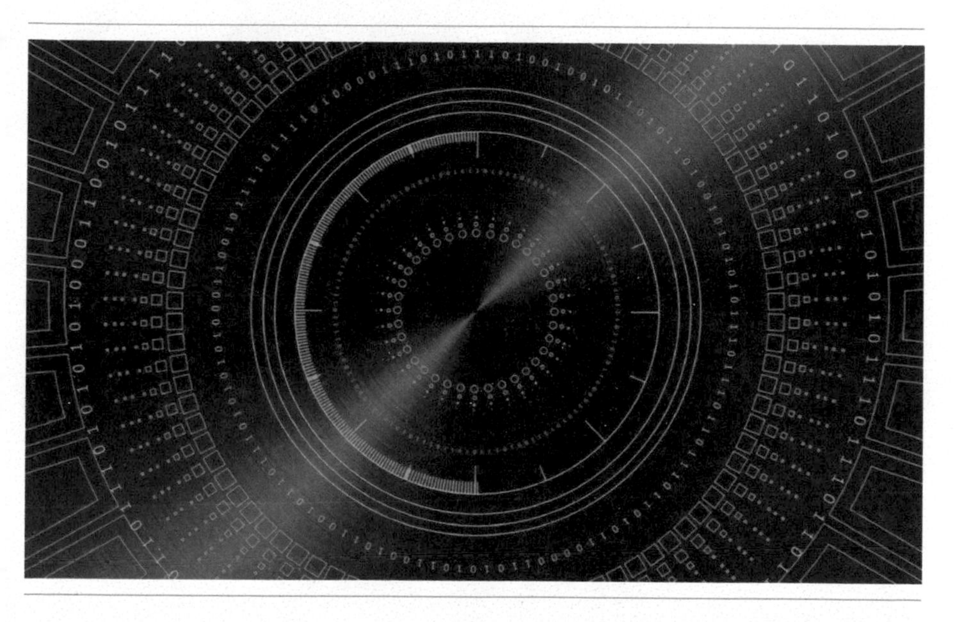

The Internet of things (IoT) – a universe of connected things pro- viding key physical data and further processing of that data in the cloud to deliver business insights – presents a huge opportunity for many players in all businesses and industries. Many companies are organizing themselves to focus on IoT and the connectivity of their future products and services. For the IoT industry to thrive, there are three categories of challenges to overcome, and this is true for any

Figure 6.1: Three major challenges facing IoT.

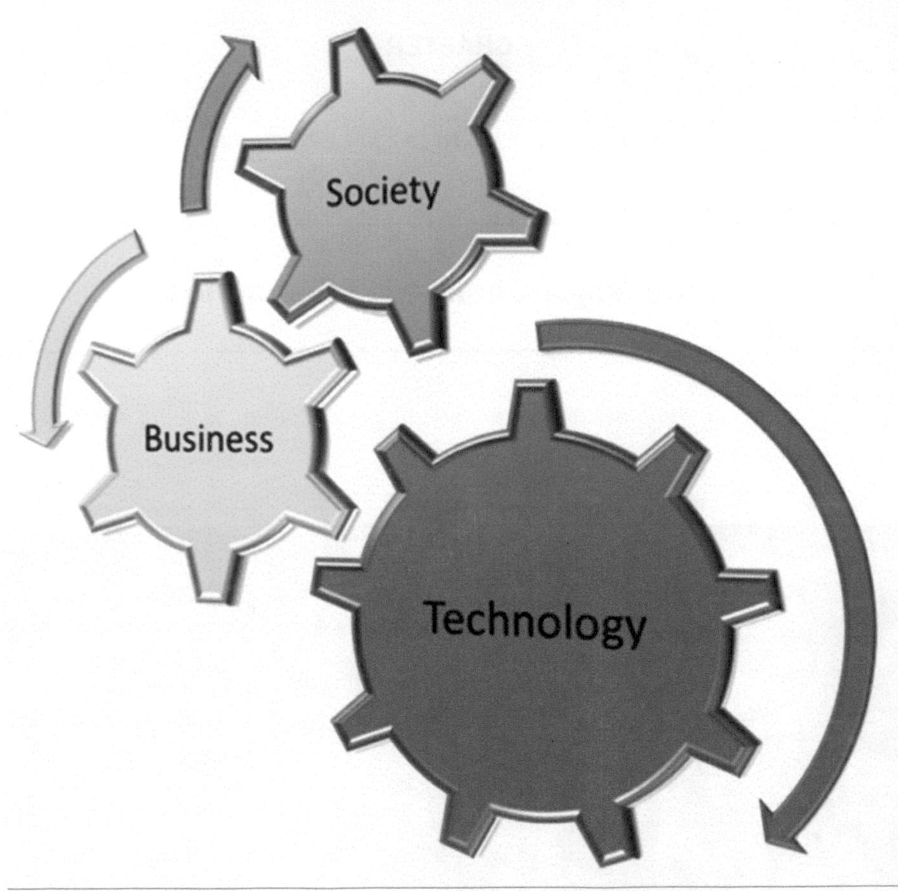

new trend in technology and not only IoT (Figure 6.1): technology, business and society [18, 19, 20].

6.1 Technology

This part covers all technologies needed to make IoT systems function smoothly as a standalone solution or part of existing systems, and that is not an easy mission – there are many technological challenges (Figure 6.2), including security, connectivity, compatibility and longevity, standards and intelligent analysis, and actions [21].

Figure 6.2: Technological challenges facing IoT.

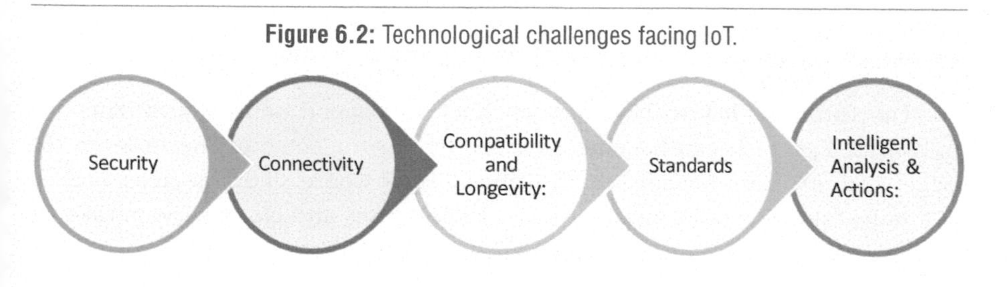

6.2 Technological Challenges

Security: IoT has already turned into a serious security concern that has drawn the attention of prominent tech firms and government agencies across the world. The hacking of baby monitors, smart fridges, thermostats, drug infusion pumps, cameras, and even the radio in your car are signifying a security nightmare being caused by the future of IoT. So many new nodes being added to networks and the Internet will provide malicious actors with innumerable attack vectors and possibilities to carry out their evil deeds, especially since a considerable number of them suffer from security holes.

The more important shift in security will come from the fact that IoT will become more ingrained in our lives. Concerns will no longer be limited to the protection of sensitive information and assets. Our very lives and health can become the target of IoT hack attacks [18].

There are many reasons behind the state of insecurity in IoT. Some of it has to do with the industry being in its "gold rush" state, where every vendor is hastily seeking to dish out the next innovative connected gadget before competitors do. Under such circumstances, functionality becomes the main focus and security takes a back seat.

Connectivity: Connecting so many devices will be one of the biggest challenges of the future of IoT, and it will defy the very structure of current communication models and the underlying technologies [19]. At present we rely on the centralized, server/client paradigm to authenticate, authorize, and connect different nodes in a network.

This model is sufficient for current IoT ecosystems, where tens, hundreds or even thousands of devices are involved. But when networks grow to join billions and hundreds of billions of devices, centralized systems will turn into a bottleneck. Such systems will require huge investments and spending in maintaining cloud servers that can handle such large amounts of information exchange, and entire systems can go down if the server becomes unavailable.

The future of IoT will very much have to depend on decentralizing IoT networks. Part of it can become possible by moving some of the tasks to the edge, such as using fog/edge computing models where smart devices such as IoT hubs take charge of mission-critical operations and cloud servers take on data gathering and analytical responsibilities [22].

Other solutions involve the use of peer-to-peer communications, where devices identify and authenticate each other directly and exchange information without the involvement of a broker. Networks will be created in meshes with no single point of failure. This model will have its own set of challenges, especially from a security perspective, but these challenges can be met with some of the emerging IoT technologies such as Blockchain [23].

Compatibility and longevity: IoT is growing in many different directions, with many different technologies competing to become the standard. This will cause difficulties and require the deployment of extra hardware and software when connecting devices.

Other compatibility issues stem from non-unified cloud services, lack of standardized M2M protocols and diversities in firmware and operating systems among IoT devices.

Some of these technologies will eventually become obsolete in the next few years, effectively rendering the devices implementing them useless. This is especially important, since in contrast to generic computing devices, which have a life span of a few years, IoT appliances (such as smart fridges or TVs) tend to remain in service for much longer, and should be able to function even if their manufacturer goes out of service.

Standards: *Technology standards*, which include network protocols, communication protocols and data-aggregation standards are the sum of all activities of handling, processing and storing the data collected from the sensors [20]. This aggregation increases the value of data by increasing *the scale, scope, and frequency* of data available for analysis.

Challenges facing the adoptions of standards within IoT

- Standard for handling unstructured data: Structured data are stored in relational databases and queried through SQL, for example. Unstructured data are stored in different types of NoSQL databases without a standard querying approach.
- Technical skills to leverage newer aggregation tools: Companies that are keen on leveraging big data tools often face a shortage of talent to plan, execute, and maintain systems.

Intelligent analysis and actions: The last stage in IoT implementation is extracting insights from data for analysis, where analysis is driven by *cognitive technologies* and the accompanying models that facilitate the use of cognitive technologies.

Factors driving adoption intelligent analytics within the IoT

- Artificial intelligence models can be improved with large data sets that are more readily available than ever before, thanks to the lower storage costs.
- Growth in crowdsourcing and open-source analytics software: Cloud-based crowdsourcing services are leading to new algorithms and improvements in existing ones at an unprecedented rate.
- Real-time data processing and analysis: Analytics tools such as complex event processing (CEP) enable processing and analysis of data on a real-time or a near-real-time basis, driving timely decision-making and action.

Challenges facing the adoptions of intelligent analytics within IoT

- Inaccurate analysis due to flaws in the data and/or model: A lack of data or presence of outliers may lead to false positives or false negatives, thus exposing various algorithmic limitations.
- Legacy systems' ability to analyze unstructured data: Legacy systems are well suited to handle structured data; unfortunately, most IoT/business interactions generate unstructured data.
- Legacy systems' ability to manage real-time data: Traditional analytics software generally works on batch-oriented processing, wherein all the data are loaded in a batch and then analyzed.

The second phase of this stage is intelligent actions, which can be expressed as M2M and M2H interfaces, for example, with all the advancement in UI and UX technologies.

Factors driving adoption of intelligent actions within the IoT

- Lower machine prices
- Improved machine functionality
- Machines "influencing" human actions through behavioral- science rationale
- Deep learning tools.

Challenges facing the adoption of intelligent actions within IoT

- Machines' actions in unpredictable situations
- Information security and privacy
- Machine interoperability
- Mean-reverting human behaviors
- Slow adoption of new technologies.

6.3 Business

The bottom line is a big motivation for starting, investing in and operating any business. Without a sound and solid business model for IoT, we will have another bubble. This model must satisfy all the requirements for all kinds of e-commerce: vertical markets, horizontal markets and consumer markets. But this category is always a victim of regulatory and legal scrutiny.

End-to-end solution providers operating in vertical industries and delivering services using cloud analytics will be the most successful at monetizing a large portion of the value in IoT. While many IoT applications may attract modest revenue, some can attract more. For little burden on the existing communication infrastructure, operators have the potential to open up a significant source of new revenue using IoT technologies.

IoT can be divided into the following three categories, based on usage and clients base:

- **Consumer IoT** includes connected devices such as smart cars, phones, watches, laptops, connected appliances and entertainment systems.
- **Commercial IoT** includes things like inventory controls, device trackers and connected medical devices.
- **Industrial IoT** covers such things as connected electric meters, wastewater systems, flow gauges, pipeline monitors, manufacturing robots and other types of connected industrial devices and systems.

Categories of IoT

Clearly, it is important to understand the value chain and business model for the IoT applications for each category of IoT (Figure 6.3).

6.4 Society

Understanding IoT from the customer's and regulator's prospective is not an easy task for the following reasons:

- Customer demands and requirements change constantly.
- New uses for devices – as well as new devices – sprout and grow at breakneck speeds.
- Inventing and reintegrating must-have features and capabilities is expensive and takes time and resources.

Figure 6.3: Categories of IoT.

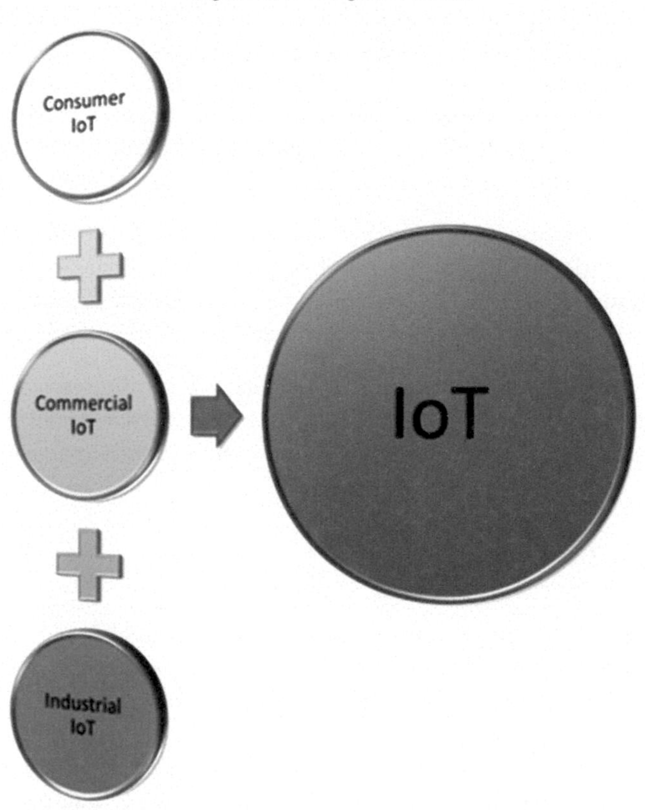

- The uses for IoT technology are expanding and changing – often in uncharted waters.
- Consumer confidence: Each of these problems could put a dent in consumers' desire to purchase connected products, which would prevent the IoT from fulfilling its true potential.
- Lack of understanding or education by consumers of best practices for IoT devices security to help in improving privacy, for example, change default passwords of IoT devices.

6.5 Privacy

The IoT creates unique challenges to privacy, many that go beyond the data privacy issues that currently exist. Much of this stems from integrating devices into our environments without us consciously using them.

This is becoming more prevalent in consumer devices, such as tracking devices for phones and cars as well as smart televisions. In terms of the latter, voice recognition or vision features are being integrated that can continuously listen to conversations or watch for activity and selectively transmit that data to a cloud service for processing, which sometimes includes a third party. The collection of this information exposes legal and regulatory challenges facing data protection and privacy law.

In addition, many IoT scenarios involve device deployments and data collection activities with multinational or global scope that cross social and cultural boundaries. What will that mean for the development of a broadly applicable privacy protection model for the IoT?

In order to realize the opportunities of the IoT, strategies will need to be developed to respect individual privacy choices across a broad spectrum of expectations, while still fostering innovation in new technologies and services.

6.6 Regulatory Standards

Regulatory standards for data markets are missing especially for data brokers – companies that sell data collected from various sources. Even though data appear to be the currency of the IoT, there is a lack of transparency about who gets access to data and how those data are used to develop products or services and are sold to advertisers and third parties. There is a need for clear guidelines on the retention, use and security of the data including metadata (the data that describe other data).

CHAPTER

7

IoT Implementation and Challenges

The Internet of things (IoT) is the network of physical objects – devices, vehicles, buildings and other items, which are embedded with electronics, software, sensors and network connectivity, which enables these objects to collect and exchange data. Implementing this concept is not an easy task by any measure for many reasons, including the complex nature of the different components of the ecosystem of IoT.

Figure 7.1: Components of IoT implementation.

To understand the gravity of this task, we will explain all the five components of IoT implementation (Figure 7.1).

7.1 Components of IoT Implementation

- Sensors
- Networks
- Standards
- Intelligent analysis
- Intelligent actions.

7.1.1 Sensors

According to IEEE, sensors can be defined as an electronic device that produces electrical, optical or digital data derived from a physical condition or event. Data produced from sensors is then electronically transformed, by another device, into information (output) that is useful in decision-making done by "intelligent" devices or individuals (people) [24].

Types of sensors: Active sensors and passive sensors.

The selection of sensors is greatly impacted by many factors, including:

- Purpose (temperature, motion, bio, etc.)
- Accuracy
- Reliability
- Range
- Resolution
- Level of intelligence (dealing with noise and interference).

The driving forces for using sensors in IoT today are new trends in technology that made sensors cheaper, smarter and smaller (Figure 7.2).

Figure 7.2: New trends of sensors.

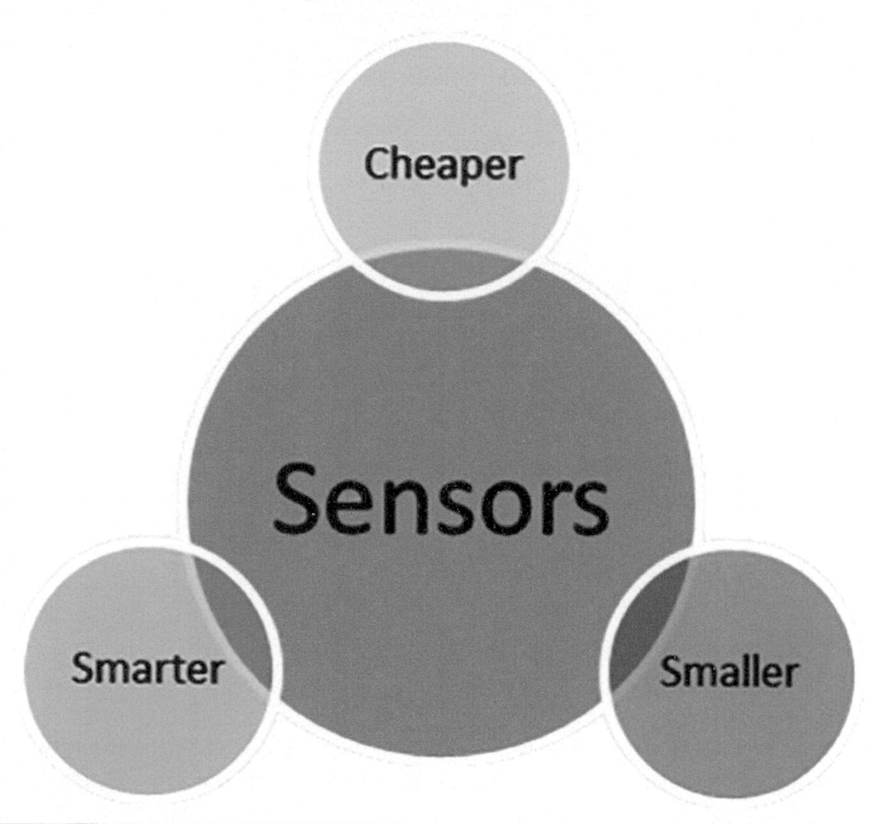

Challenges facing IoT sensors:

- Power consumption
- Security
- Interoperability.

7.1.2 Networks

The second step of this implementation is to transmit the signals collected by sensors over networks with all the different components of a typical network including routers, bridges in different topologies, including LAN, MAN, and WAN. Connecting the different parts of networks to the sensors can be done by different technologies, including Wi-Fi, Bluetooth, low power Wi-Fi, Wi-Max, regular Ethernet, Long Term Evolution (LTE) and the recent promising technology of Li-Fi (using light as a medium of communication between the different parts of a typical network including sensors).

The driving forces for widespread network adoption in IoT can be summarized as follows:

- High data rate
- Low prices of data usage
- Virtualization (X-Defined Network trends)
- XaaS concept (SaaS, PaaS and IaaS)
- IPv6 deployment,

Challenges facing network implementation in IoT

- The enormous growth in the number of connected devices
- Availability of networks coverage
- Security
- Power consumption.

7.1.3 Standards

The third stage in the implementation process involves all activities of handling, processing and storing the data collected from the sensors. This aggregation increases the value of data by increasing *the scale, scope and frequency* of data available for analysis but aggregation is only achieved through the use of various standards depending on the IoT application used.

Types of Standards

Two types of standards are relevant for the aggregation process: *technology standards* (including network protocols, communication protocols and data aggregation standards) and *regulatory standards* (related to security and privacy of data, among other issues).

Technology Standards

- Network protocols (e.g., Wi-Fi)
- Communications protocols (e.g., HTTP)
- Data aggregation standards (e.g., extraction, transformation, loading (ETL)).

Regulatory Standards

Set and administrated by government agencies like FTC, for example, Fair Information Practice Principles (FIPP) and US Health Insurance Portability and Accountability Act (HIPAA), just to mention a few.

Challenges facing the adoptions of standards within IoT

- **Standard for handling unstructured data:** Structured data are stored in relational databases and queried through SQL. Unstructured data are stored in different types of NoSQL databases without a standard querying approach.
- **Security and privacy issues:** There is a need for clear guidelines on the retention, use and security of the data as well as metadata (the data that describes other data).
- **Regulatory standards for data markets:** Data brokers are companies that sell data collected from various sources. Even though data appear to be the currency of the IoT, there is a lack of transparency about who gets access to data and how those data are used to develop products or services and sold to advertisers and third parties.
- **Technical skills to leverage newer aggregation tools:** Companies that are keen on leveraging Big Data tools often face a shortage of talent to plan, execute and maintain systems.

7.1.4 Intelligent Analysis

The fourth stage in IoT implementation is extracting insight from data for analysis, Analysis is driven by *cognitive technologies* and the accompanying models that facilitate the use of cognitive technologies. With advances in cognitive technologies' ability to process varied forms of information, vision and voice have also become usable. Below is a list of selected cognitive technologies that are experiencing increasing adoption and can be deployed for predictive and prescriptive analytics:

- **Computer vision** refers to a computer's ability to identify objects, scenes and activities in images.
- **Natural-language processing** refers to a computer's ability to work with text the way humans do, extracting meaning from text or even generating text.
- **Speech recognition** focuses on accurately transcribing human speech.

Factors driving adoption intelligent analytics within the IoT

- **Artificial intelligence models** can be improved with large data sets that are more readily available than ever before, thanks to the lower storage costs.

- **Growth in crowdsourcing and open-source analytics software:** Cloud-based crowdsourcing services are leading to new algorithms and improvements in existing ones at an unprecedented rate.
- **Real-time data processing and analysis:** Analytics tools such as complex event processing (CEP) enable processing and analysis of data on a real-time or a near-real-time basis, driving timely decision-making and action.

Challenges facing the adoptions of intelligent analytics within IoT

- **Inaccurate analysis due to flaws in the data and/or model:** A lack of data or presence of outliers may lead to false positives or false negatives, thus exposing various algorithmic limitations.
- **Legacy systems' ability to analyze unstructured data:** Legacy systems are well suited to handle structured data; unfortunately, most IoT/business interactions generate unstructured data.
- **Legacy systems' ability to manage real-time data:** Traditional analytics software generally works on batch-oriented processing, wherein all the data are loaded in a batch and then analyzed.

7.1.5 Intelligent Actions

Intelligent actions can be expressed as M2M and M2H interface, for example, with all the advancement in UI and UX technologies.

Factors driving adoption of intelligent actions within the IoT

- Lower machine prices
- Improved machine functionality
- Machines "influencing" human actions through behavioral- science rationale
- Deep learning tools.

Challenges facing the adoption of intelligent actions within IoT

- Machines' actions in unpredictable situations
- Information security and privacy
- Machine interoperability
- Mean-reverting human behaviors
- Slow adoption of new technologies.

The IoT is an ecosystem of ever-increasing complexity, it is the next wave of innovation that will humanize every object in our lives, which is the next level to automating every object in our life. Convergence of technologies will make IoT implementation much easier and faster, which in turn will improve many aspects of our life at home and at work and in between [25].

CHAPTER

8

IoT Standardization and Implementation Challenges

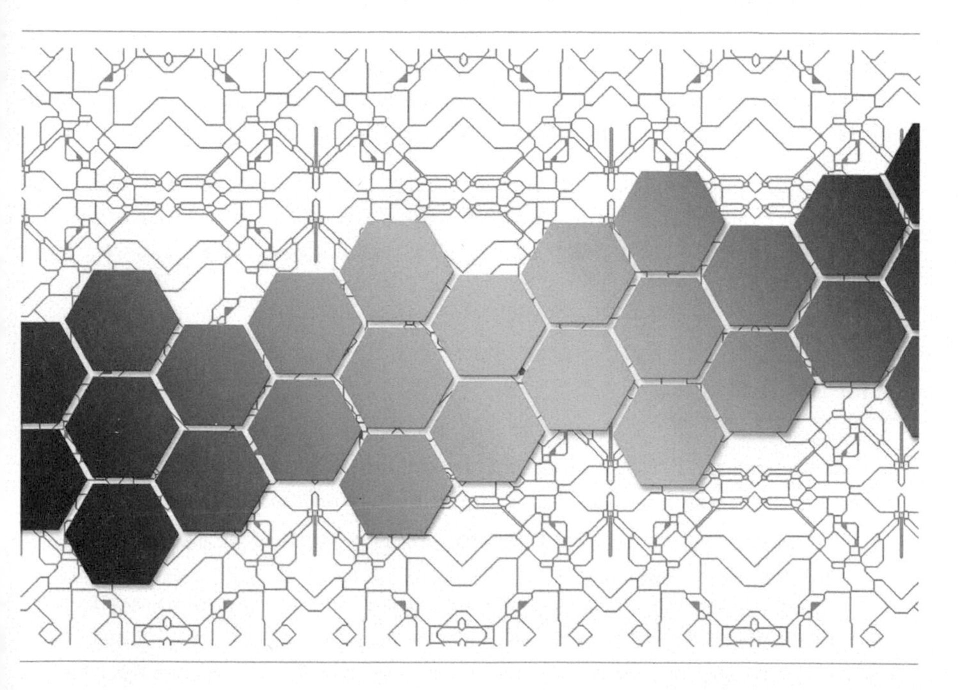

The rapid evolution of the IoT market has caused an explosion in the number and variety of IoT solutions. Additionally, large amounts of funding are being deployed at IoT startups. Consequently, the focus of the industry has been on manufacturing and producing the right types of hardware to enable those solutions. In the current model, most IoT solution providers have been building all components of the stack, from the hardware devices to the relevant cloud services or as they would like to name it, "IoT solutions". As a result, there is a lack of consistency and standards across the cloud services used by the different IoT solutions.

Figure 8.1: Hurdles facing IoT standardization.

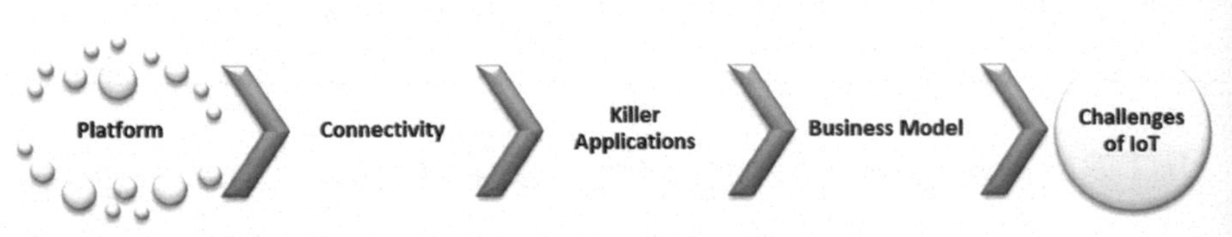

As the industry evolves, the need for a standard model to perform common IoT backend tasks, such as processing, storage and firmware updates, is becoming more relevant. In that new model, we are likely to see different IoT solutions work with common backend services, which will guarantee levels of interoperability, portability and manageability that are almost impossible to achieve with the current generation of IoT solutions.

Creating that model will never be an easy task by any level of imagination. There are hurdles and challenges facing the standardization and implementation of IoT solutions (Figure 8.1), and that model needs to overcome all of them [29].

8.1 IoT Standardization

The hurdles facing IoT standardization can be divided into four categories: platform, connectivity, business model and killer applications.

- **Platform:** This part includes the form and design of the products (UI/UX), analytics tools used to deal with the massive data streaming from all products in a secure way and scalability, which means wide adoption of protocols like IPv6 in all vertical and horizontal markets is needed.
- **Connectivity:** This phase includes all parts of the consumer's routine, from using wearables, smart cars, smart homes and, in the big scheme, smart cities. From the business perspective, we have connectivity using IIoT (industrial Internet of things), where M2M communications dominate the field.
- **Killer applications:** In this category, there are three functions needed to have killer applications: control "things", collect "data" and analyze "data". IoT needs killer applications to drive the business model using a unified platform.

- **Business model:** The bottom line is a big motivation for starting, investing in, and operating any business. Without a sound and solid business model for IoT, we will have another bubble, and this model must satisfy all the requirements for all kinds of e- commerce: vertical markets, horizontal markets and consumer markets. But this category is always a victim of regulatory and legal scrutiny.

All four categories are inter-related, and you need to make all of them work. Missing one will break that model and stall the standardization process. A lot of work is needed in this process, and many companies are involved in each of one of the categories. Bringing them to the table to agree on a unifying model will be a daunting task.

8.2 IoT Implementation

The second part of the model is IoT implementations; implementing IoT is not an easy process by any measure for many reasons, including the complex nature of the different components of the ecosystem of IoT. To understand the significance of this process, we will explore all the *five* components of IoT implementation: sensors, networks, standards, intelligent analysis and intelligent actions (Figure 8.2).

8.2.1 Sensors

There two types of sensors: active sensors and passive sensors. The driving forces for using sensors in IoT today are new trends in technology that made sensors cheaper, smarter and smaller. But the challenges facing IoT sensors are power consumption, security and interoperability.

8.2.2 Networks

The second component of IoT implantation is to transmit the signals collected by sensors over networks with all the different components of a typical network including routers and bridges in different topologies. Connecting the different parts of networks to the sensors can be done by different technologies, including Wi-Fi, Bluetooth, low power Wi-Fi, Wi-Max, regular Ethernet, Long Term Evolution (LTE) and the recent promising technology of Li-Fi (using light

Figure 8.2: Components of IoT implementations.

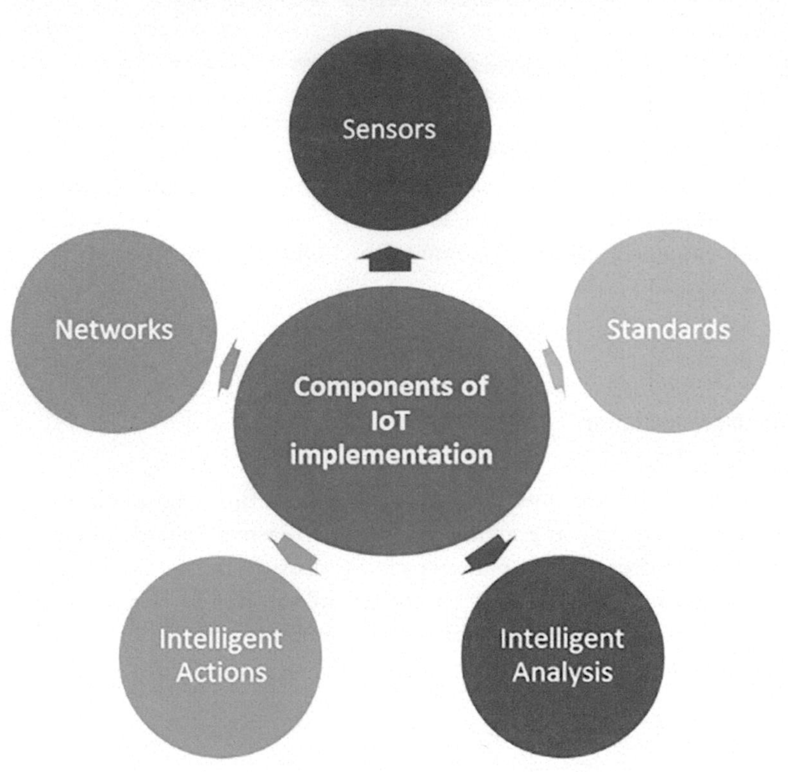

as a medium of communication between the different parts of a typical network including sensors) [27].

The driving forces for widespread network adoption in IoT are high data rate, low prices of data usage, virtualization (X-Defined Network trends), XaaS concept (SaaS, PaaS and IaaS) and IPv6 deployment. But the challenges facing network implementation in IoT are the enormous growth in a number of connected devices, availability of networks coverage, security and power consumption.

8.2.3 Standards

The third stage in the implementation process includes the sum of all activities of handling, processing and storing the data collected from the sensors. This aggregation increases the value of data by increasing the scale, scope and frequency of data available for analysis, but aggregation is only achieved through the use of various standards depending on the IoT application used.

There are two types of standards relevant for the aggregation process: *technology standards* (including network protocols, communication protocols and data aggregation standards) and *regulatory standards* (related to security and privacy of data, among other issues). Challenges facing the adoptions of standards within IoT are standard for handling unstructured data, security and privacy issues in addition to regulatory standards for data markets [26].

8.2.4 Intelligent Analysis

The fourth stage in IoT implementation is extracting insight from data for analysis. IoT analysis is driven by *cognitive technologies* and the accompanying models that facilitate the use of cognitive technologies. With advances in cognitive technologies, the ability to process varied forms of information, vision and voice has also become usable and opened the doors for an in-depth understanding of the non-stop streams of real-time data. Factors driving adoption intelligent analytics within the IoT are: artificial intelligence models, growth in crowdsourcing and open-source analytics software, real-time data processing and analysis. Challenges facing the adoption of analytics within IoT are: inaccurate analysis due to flaws in the data and/or model, legacy systems' ability to analyze unstructured data and legacy systems' ability to manage real-time data [35].

8.2.5 Intelligent Actions

Intelligent actions can be expressed as M2M (machine to machine) and M2H (machine to human) interfaces, for example, with all the advancement in UI and UX technologies. Factors driving adoption of intelligent actions within the IoT are: lower machine prices, improved machine functionality, machines "influencing" human actions through behavioral-science rationale and deep learning tools. Challenges facing the adoption of intelligent actions within IoT are: machines' actions in unpredictable situations, information security and privacy, machine interoperability, mean-reverting human behaviors, and slow adoption of new technologies.

8.3 The Road Ahead

The IoT is an ecosystem of ever-increasing complexity. It is the next wave of innovation that will humanize every object in our lives, and is the next level to automating every object in our lives and convergence of technologies that will make IoT implementation much easier and faster, which in turn will improve many aspects of our lives at home and at work and in between. From refrigerators to parking spaces to houses, IoT is bringing more and more things into the digital fold every day, which will likely make IoT a multitrillion-dollar industry in the near future. One possible outcome of successful standardization of IoT is the implementation of "IoT as a service" technology if that service is offered and used in the same way we use other flavors of "as a service" technologies. Today, the possibilities of applications in real life are unlimited. But we have a long way to achieve that dream, and we need to overcome many obstacles and barriers on two fronts, consumers and businesses, before we can harvest the fruits of such technology [28, 30].

CHAPTER

9

Challenges Facing IoT Analytics Success

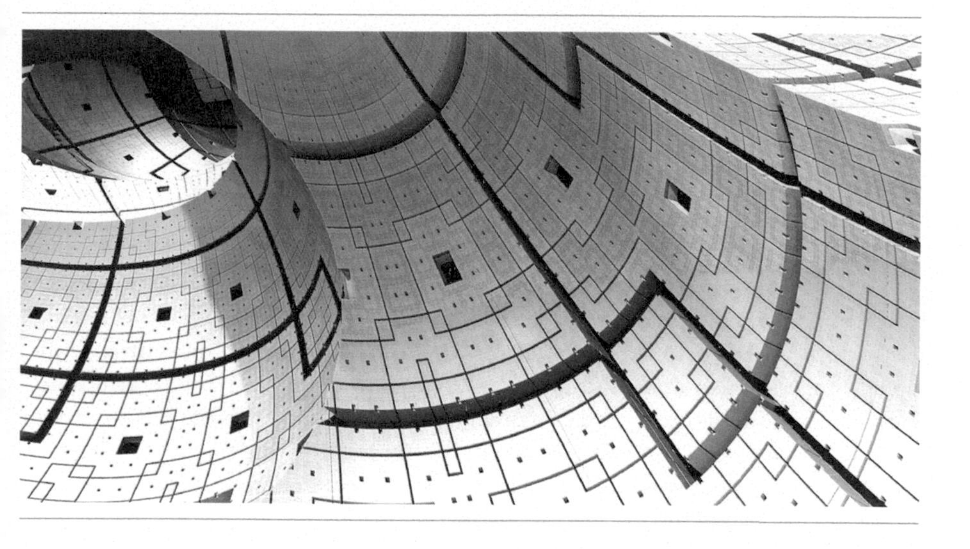

The Internet of things (IoT) is an ecosystem of ever-increasing complexity; it is the next wave of innovation that will humanize every object in our life. IoT is bringing more and more devices (things) into the digital fold every day, which will likely make IoT a multitrillion- dollar industry in the near future. To understand the scale of interest in the IoT, just check how many conferences, articles and studies have been conducted about IoT recently, including a recent well-written article about the IoT future by SAP listing 21 experts' insights into the IoT future. This interest hit fever pitch point last year as many companies saw big opportunities and believed that IoT holds the promise to expand and improve businesses processes and accelerate growth.

However, the rapid evolution of the IoT market has caused an explosion in the number and variety of IoT solutions, which has created real challenges as the industry evolves, mainly the urgent need for a reliable IoT model to perform common tasks such as sensing, processing, storage, and communication. Developing that model will never be an easy task by any stretch of the imagination; there are many hurdles and challenges facing a real reliable IoT model.

One of the crucial functions of using IoT solutions is to take advantage of IoT analytics to exploit the information collected by “things” in many ways, for example, to understand customer behavior, to deliver services, to improve products, and to identify and intercept business moments. IoT demands new analytic approaches as data volumes increase through 2022 to astronomical levels, and the needs of the IoT analytics may diverge further from traditional analytics.

There are many challenges facing IoT analytics (Figure 9.1), including data structures, combing multi data formats, the need to balance scale and speed, analytics at the edge, and IoT analytics and AI.

9.1 Data Structures

Most sensors send out data with a time stamp and most of the data is “boring” with nothing happening for much of the time. However, once in a while, something serious happens and needs to be attended to. While static alerts based on thresholds are a good starting point for analyzing this data, they cannot help us advance to diagnostic or predictive or prescriptive phases. There may be relationships between data pieces collected at specific intervals of time, in other words, classic time series challenges.

9.2 Combining Multiple Data Formats

While time series data have established techniques and processes for handling, the insights that would really matter cannot come from sensor data alone. There are usually strong correlations between sensor data and other unstructured data. For example, a series of control unit fault codes may result in a specific service action that is recorded by a mechanic. Similarly, a set of temperature

Figure 9.1: Challenges facing IoT analytics.

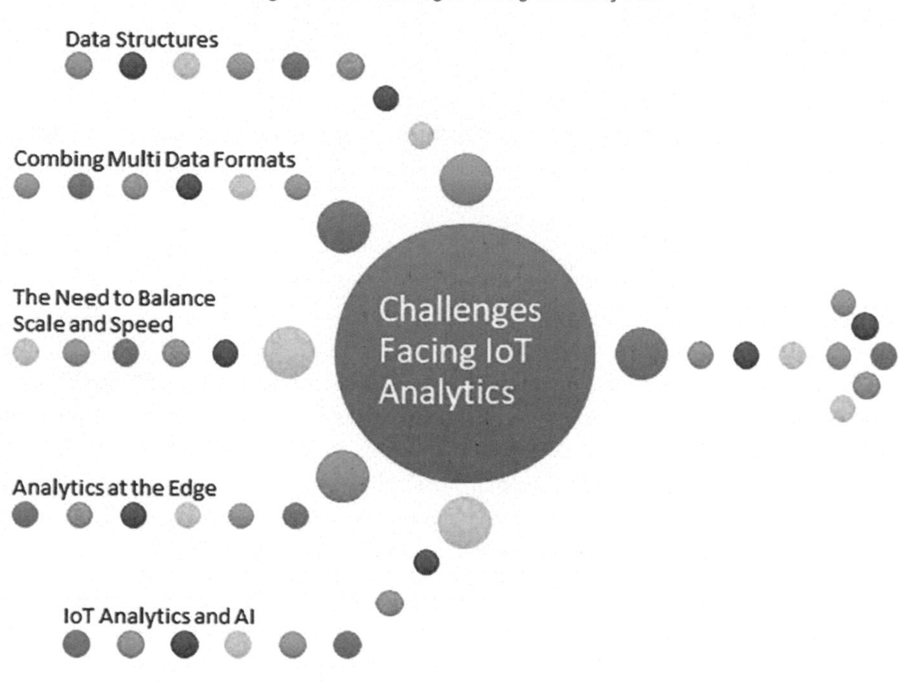

readings may be accompanied by a sudden change in the macroscopic shape of a part that can be captured by an image or change in the audible frequency of a spinning shaft. We would need to develop techniques where structured data must be effectively combined with unstructured data or what we call "dark data" [31].

9.3 The Need to Balance Scale and Speed

Most of the serious analysis for IoT will happen in the cloud, a data center, or more likely a hybrid cloud and server-based environment. This is because, despite the elasticity and scalability of the cloud, it may not be suited for scenarios requiring large amounts of data to be processed in real time. For example, moving 1 terabyte over a 10 Gbps network takes 13 minutes, which is fine for batch processing and management of historical data, but it is not practical for analyzing real-time event streams. A recent example is data transmitted by autonomous cars especially in critical situations that required a split second decision.

At the same time, because different aspects of IoT analytics may need to scale more than others, the analysis algorithm implemented should support flexibility whether the algorithm is deployed in the edge, data center, or cloud.

9.4 IoT Analytics at the Edge

IoT sensors, devices and gateways are distributed across different manufacturing floors, homes, retail stores and farm fields, to name just a few locations. Yet moving 1 terabyte of data over a 10 Mbps broadband network will take 9 days. Therefore, enterprises need to plan on how to address the projected 40% of IoT data that will be processed at the edge in just a few years' time. This is particularly true for large IoT deployments, where billions of events may stream through each second, but systems only need to know an average over time or be alerted when a trends fall outside established parameters.

The answer is to conduct some analytics on IoT devices or gateways at the edge and send aggregated results to the central system. Through such edge analytics, organizations can ensure the timely detection of important trends or aberrations while significantly reducing network traffic to improve performance.

Performing edge analytics requires very lightweight software, since IoT nodes and gateways are low-power devices with limited strength for query processing. To deal with this challenge, fog/edge computing is the champion [31].

Fog/edge computing allows computing, decision-making and action-taking to happen via IoT devices and only pushes relevant data to the cloud. Cisco coined the term "fog/edge computing" and gave a brilliant definition for it: "The fog/edge extends the cloud to be closer to the things that produce and act on IoT data. These devices, called fog/edge nodes, can be deployed anywhere with a network connection: on a factory floor, on top of a power pole, alongside a railway track, in a vehicle, or on an oil rig. Any device with computing, storage, and network connectivity can be a fog/edge node. Examples include industrial controllers, switches, routers, embedded servers, and video surveillance cameras." The major benefits of using fog/edge computing are: minimizing latency, conserving network bandwidth and addressing security concerns at all levels of the network. In addition, it operates reliably with quick decisions, collects and secures a wide range of data, moves data to the best place for processing, lowers expenses of using high computing power only when needed, uses less bandwidth, and gives better analysis and insights of local data.

Keep in mind that fog/edge computing is not a replacement for cloud computing by any measures; it works in conjunction with cloud computing, optimizing the use of available resources. But it was the product of a need to address many challenges: real-time process and action of incoming data, as well as limitation of resources like bandwidth and computing power, another factor helping fog/edge computing is the fact that it takes advantage of the distributed nature of today's virtualized IT resources. This improvement to the data-path hierarchy is enabled by the increased compute functionality that manufacturers are building into their edge routers and switches.

9.5 IoT Analytics and AI

The greatest – and as yet largely untapped – power of IoT analysis is to go beyond reacting to issues and opportunities in real time and instead prepare for them beforehand. That is why prediction is central to many IoT analytics strategies, whether to project demand, anticipate maintenance, detect fraud, predict churn or segment customers.

Artificial intelligence (AI) uses and improves current statistical models for handling prediction. AI will automatically learn underline rules, providing an attractive alternative to rules-only systems, which require professionals to author rules and evaluate their performance. When AI is applied, it provides valuable and actionable insights.

There are six types of IoT data analysis where AI can help (Figure 9.2):

1. **Data preparation:** Defining pools of data and cleaning them, which will take us to concepts like dark data and data lakes [32].
2. **Data discovery:** Finding useful data in the defined pools of data.
3. **Visualization of streaming data:** On the fly dealing with streaming data by defining, discovering, and visualizing data in smart ways to make it easy for the decision-making process to take place without delay.
4. **Time series accuracy of data:** Keeping the level of confidence in data collected high with high accuracy and integrity of data.
5. **Predictive and advance analytics:** Very important step where decisions can be made based on data collected, discovered and analyzed.
6. **Real-time geospatial and location (logistical data):** Maintaining the flow of data smoothly and under control.

Figure 9.2: AI and IoT data analysis.

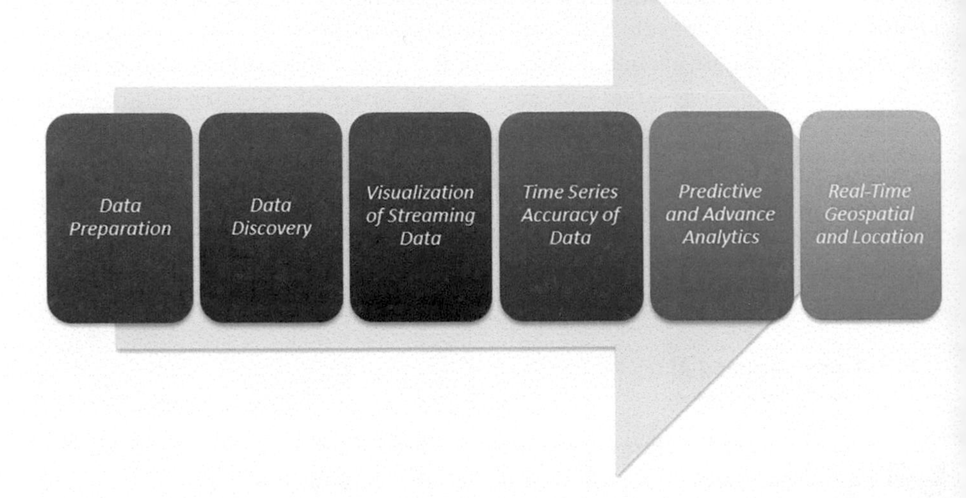

But it is not all "nice and rosy, comfy and cozy", as there are challenges in using AI in IoT [33], such as compatibility, complexity, privacy/security/safety, ethical and legal issues, and artificial stupidity. Many IoT ecosystems will emerge, and commercial and technical battles between these ecosystems will dominate areas such as the smart home, the smart city, financials and healthcare. But the real winners will be the ecosystems with better, reliable, fast and smart IoT analytics tools, after all what matters is how can we change data to insights and insights to actions and actions to profit.

CHAPTER

10

How to Secure the Internet of Things

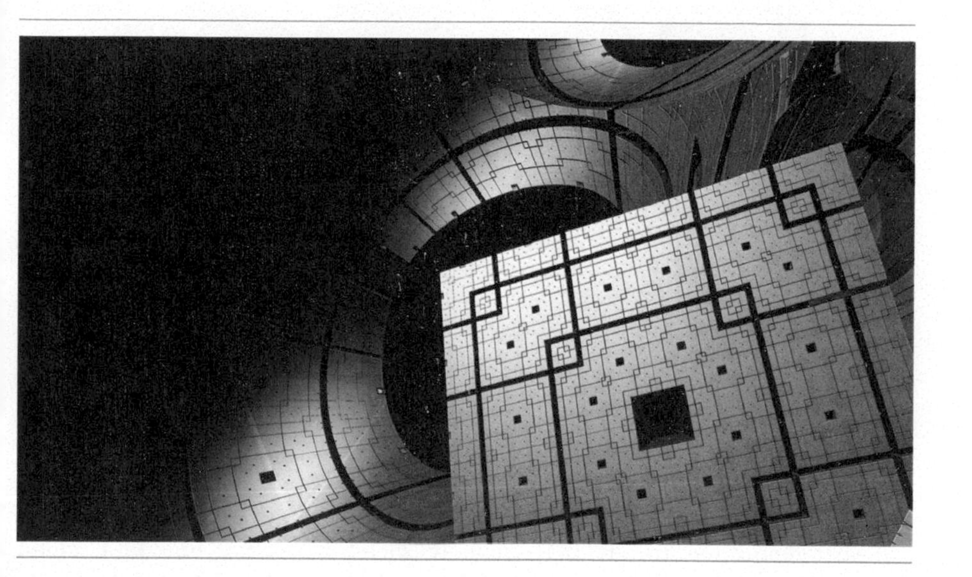

The Internet of things (IoT) as a concept is fascinating and exciting, but the key to gaining real business value from it is effective communication between all elements of the architecture so you can deploy applications faster, process and analyze data at lightning speeds, and make decisions as soon as you can.

IoT architecture can be represented by four systems (Figure 10.1):

1. **Things:** These are defined as uniquely identifiable nodes, primarily sensors that communicate without human interaction using IP connectivity.

2. **Gateways:** These act as intermediaries between things and the cloud to provide the needed Internet connectivity, security and manageability.
3. **Network infrastructure:** This is composed of routers, aggregators, gateways, repeaters and other devices that control data flow.
4. **Cloud infrastructure:** Cloud infrastructure contains large pools of virtualized servers and storage that are networked together.

Figure 10.1: IoT architecture.

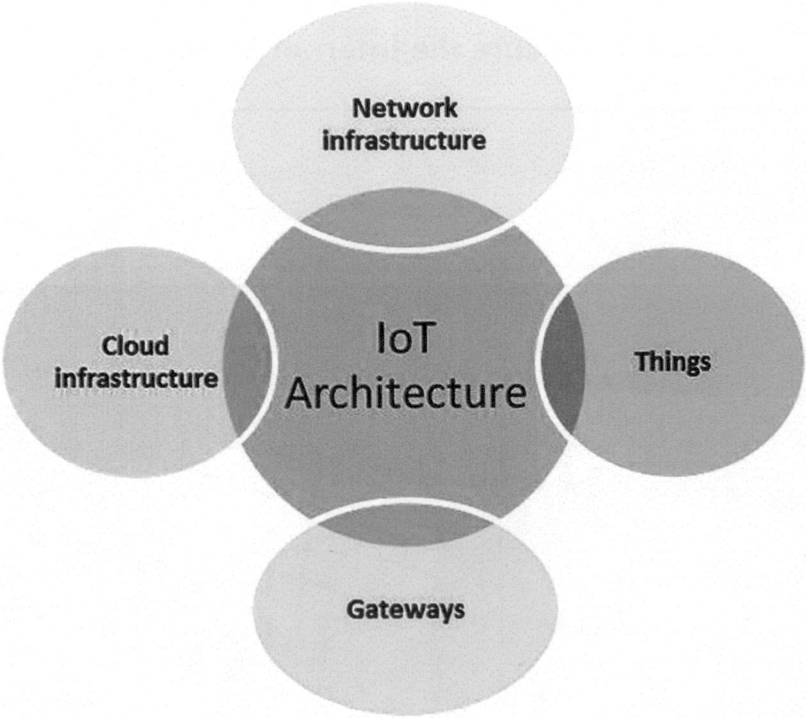

Next-generation trends [34], namely social networks, big data, cloud computing, and mobility, have made many things possible that were not just a few years ago. Add to that the convergence of global trends and events that are fueling today's technological advances and enabling innovation including:

- Efficiency and cost-reduction initiatives in key vertical markets.
- Government incentives encouraging investment in these new technologies.
- Lower manufacturing costs for smart devices.
- Reduced connectivity costs.
- More efficient wired and wireless communications.
- Expanded and affordable mobile networks.

The IoT is one big winner in this entire ecosystem; it is creating new opportunities and providing a competitive advantage for businesses in current and new markets. It touches everything – not just the data, but how, when, where, and why you collect it. The technologies that have created the IoT are not changing the Internet only, but rather changing the things connected to the internet – the devices and gateways on the edge of the network that are now able to request a service or start an action without human intervention at many levels.

Because the generation and analysis of data is so essential to the IoT, consideration must be given to protecting data throughout its life cycle. Managing information at this level is complex because data will flow across many administrative boundaries with different policies and intents. Generally, data is processed or stored on edge devices that have highly limited capabilities and are vulnerable to sophisticated attacks. Given the various technological and physical components that truly make up an IoT ecosystem, it is good to consider the IoT as a system of systems. The architecting of these systems that provide business value to organizations will often be a complex task, as enterprise architects work to design integrated solutions that include edge devices [35], applications, transports, protocols, and analytics capabilities that make up a fully functioning IoT system. This complexity introduces challenges to keep the IoT secure and ensure that a particular instance of the IoT cannot be used as a jumping off point to attack other enterprise information technology (IT) systems.

The International Data Corporation (IDC) reported that 90% of organizations that implement the IoT suffered an IoT-based breach of back-end IT systems.

10.1 Challenges to Secure IoT Deployments

Regardless of the role your business has within the IoT ecosystem – device manufacturer, solution provider, cloud provider, systems integrator or service provider – you need to know how to get the greatest benefit from this new technology that offers such highly diverse and rapidly changing opportunities.

Handling the enormous volume of existing and projected data is daunting. Managing the inevitable complexities of connecting to a seemingly unlimited list of devices is complicated. And the goal of turning the deluge of data into valuable actions seems impossible because of the many challenges. The existing security technologies will play a role in mitigating IoT risks, but they are not enough. The goal is to get data securely at the right place, at the right time and in the right format, and it is easier said than done for many reasons. The Cloud Security Alliance (CSA) [36], in a recent report, listed some of the challenges:

- Many IoT systems are poorly designed and implemented, using diverse protocols and technologies that create complex configurations.
- Lack of mature IoT technologies and business processes.
- Limited guidance for life cycle maintenance and management of IoT devices.
- The IoT introduces unique physical security concerns.
- IoT privacy concerns are complex and not always readily evident.
- Limited best practices available for IoT developers.
- There is a lack of standards for authentication and authorization of IoT edge devices.
- There are no best practices for IoT-based incident response activities.
- Audit and logging standards are not defined for IoT components.
- Restricted interfaces available IoT devices to interact with security devices and applications.
- No focus yet on identifying methods for achieving situational awareness of the security posture of an organization's IoT assets.
- Security standards, for platform configurations, involving virtualized IoT platforms supporting multi-tenancy is immature.
- Customer demands and requirements change constantly.
- New uses for devices – as well as new devices – sprout and grow at breakneck speeds.
- Inventing and reintegrating must-have features and capabilities are expensive and consume time and resources.
- The uses for IoT technology are expanding and changing – often in uncharted waters.
- Developing the embedded software that provides IoT value can be difficult and expensive.

Some real examples of threats and attack vectors that malicious actors could take advantage of are:

- Control systems, vehicles and even the human body can be accessed and manipulated causing injury or worse.
- Healthcare providers can improperly diagnose and treat patients.
- Intruders can gain physical access to homes or commercial businesses.
- Loss of vehicle control.
- Safety critical information such as warnings of a broken gas line can go unnoticed.
- Critical infrastructure damage.
- Malicious parties can steal identities and money.
- Unanticipated leakage of personal or sensitive information.
- Unauthorized tracking of people's locations, behaviors, and activities.
- Manipulation of financial transactions.
- Vandalism, theft or destruction of IoT assets.
- Ability to gain unauthorized access to IoT devices.
- Ability to impersonate IoT devices.

10.2 Dealing with the Challenges and Threats

Gartner predicted at its security and risk management summit in Mumbai, India, this year, that more than 40% of businesses will have deployed security solutions for protecting their IoT devices and services by 2022. IoT devices and services will expand the surface area for cyber-attacks on businesses, by turning physical objects that used to be offline into online assets communicating with enterprise networks. Businesses will have to respond by broadening the scope of their security strategy to include these new online devices.

Businesses will have to tailor security to each IoT deployment according to the unique capabilities of the devices involved and the risks associated with the networks connected to those devices. BI Intelligence expects spending on solutions to secure IoT devices and systems to increase five-fold over the next 4 years.

10.3 The Optimum Platform

Developing solutions for the IoT requires unprecedented collaboration, coordination, and connectivity for each piece in the system, and throughout the system as a whole. All devices must work together and be integrated with all other devices, and all devices must communicate and interact seamlessly with connected systems and infrastructures. It is possible, but it can be expensive, time consuming, and difficult.

The optimum platform for IoT can:

- Acquire and manage data to create a standards-based, scalable and secure platform.
- Integrate and secure data to reduce cost and complexity while protecting your investment.
- Analyze data and act by extracting business value from data and then acting on it.

10.4 Last Word

Security needs to be built in as the foundation of IoT systems, with rigorous validity checks, authentication, and data verification, and all the data needs to be encrypted. At the application level, software development organizations need to be better at writing code that is stable, resilient and trustworthy, with better code development standards, training, threat analysis and testing. As systems interact with each other, it is essential to have an agreed interoperability standard, which is safe and valid. Without a solid bottom–top structure, we will create more threats with every device added to the IoT. What we need is a secure and safe IoT with privacy-protected, tough trade-off, but it is not impossible.

References

[1] https://www.goldmansachs.com/insights/pages/internet-of-things/

[2] https://openconnectivity.org/developer/reference-implementation/alljoyn

[3] https://www.accenture.com/us-en/new-applied-now

[4] https://www.zdnet.com/article/welcome-to-the-dystopian-internet-of-things-powered-by-and-starring-you/

[5] https://www.bbvaopenmind.com/en/the-internet-of-everything-ioe/

[6] http://www.eweek.com/blogs/first-read/cisco-ceo-internet-of-everything-will-be-worth-19-trillion

[7] http://www.eweek.com/security/cisco-sponsors-300-000-internet-of-things-security-challenge

[8] https://www.ndtv.com/business/internet-will-disappear-google-boss-tells-davos-731773

[9] https://www.csoonline.com/article/2134066/mobile-security/what-the-internet-of-things-means-for-security.html

[10] https://security.vt.edu/about/facultystaff/randymarchany.html

[11] https://www.owasp.org/index.php/MainPage

[12] https://www.linkedin.com/pulse/industrial-internet-things-iiot-challenges-benefits-ahmed-banafa/?trk=mp-reader-card

[13] https://www.bbvaopenmind.com/en/iot-securityprivacy-and-safety/

[14] https://www.cisco.com/c/en/us/solutions/internet-of-things/smart-city-infrastructure-guide.html#~stickynav=1

[15] https://www.linkedin.com/pulse/20140520183722-246665791-fog-computing/?trk=mp-reader-card

[16] http://www.mouser.com/pdfdocs/IOTEXECBRIEFWP.PDF

[17] https://www.linkedin.com/pulse/20140312180810-246665791-the-future-of-big-data-and-analytics/?trk=mp-reader-card

[18] http://www.microwavejournal.com/articles/27690-addressing-the-challenges-facing-iot-adoption

[19] https://blog.apnic.net/2015/10/20/5-challenges-of-the-internet-of-things/

[20] https://www.sitepoint.com/4-major-technical-challenges-facing-iot-developers/

[21] https://www.linkedin.com/pulse/iot-implementation-challenges-ahmed-banafa?trk=mp-author-card

[22] https://www.bbvaopenmind.com/en/why-iot-needs-fog-computing/

[23] http://iot.ieee.org/newsletter/january-2017/iot-and-blockchain-convergence-benefits-and-challenges.html

[24] https://ieeexplore.ieee.org/document/5739775/

[25] https://www2.deloitte.com/insights/us/en.html

[26] http://www.dbta.com/BigDataQuarterly/Articles/10-Predictions-for-the-Future-of-IoT-109996.aspx

[27] https://campustechnology.com/articles/2016/02/25/security-tops-list-of-trends-that-will-impact-the-internet-of-things.aspx

[28] http://dupress.com/

[29] https://datafloq.com/read/iot-standardization-and-implementation-challenges/2164

[30] https://www.bbvaopenmind.com/en/what-is-next-for-iot/

[31] https://www.bbvaopenmind.com/en/understanding-dark-data/

[32] https://www.bbvaopenmind.com/en/data-lake-an-opportunity-or-a-dream-for-big-data/

[33] https://datafloq.com/read/why-ai-is-the-catalyst-of-iot/3046

[34] https://www.bbvaopenmind.com/en/cloud-computing-big-data-and-mobility-2015-tech-trends/

[35] https://datafloq.com/read/fog-computing-vital-successful-internet-of-things/1166

[36] https://cloudsecurityalliance.org/

Index